WIRING SIMPLIFIED
Based on the 1999 National Electrical Code®

by

H. P. Richter

as revised by

W. Creighton Schwan

39th edition

PARK PUBLISHING, INC.
Minneapolis, Minnesota
Somerset, Wisconsin

PARK PUBLISHING, INC.

350 Main Street, P.O. Box 387
Somerset, Wisconsin 54025-0387

Printed in the United States of America

ISBN 0-9603294-7-1

Warnings and Disclaimer

This book provides useful general instruction, but we cannot anticipate all of your working conditions or the characteristics of your materials and tools. For your safety, and the safety of others, you must use caution, care, and good judgment when following the procedures described in this book. You must adapt the instructions to your particular situation. Consider your own skill level and follow the instructions and safety precautions and warnings associated with the various tools and materials shown. Electricity and certain tools can be dangerous and can cause injury.

Information has been obtained from sources believed to be reliable, but its accuracy and completeness are not guaranteed. The publisher and author disclaim liability, whether for personal injury or property damage, arising out of typographical error or the use or misuse of this publication regardless of the fault or negligence of the publisher or author. The instructions in this book conform to the 1999 National Electrical Code®. Consult your local authorities for information on building permits, codes, and other laws as they apply to your project.

Publisher's Cataloging-in-Publication

Richter, H. P. (Herbert P.), 1900–1978.
    Wiring simplified : based on the 1999
National Electrical Code / by H. P. Richter, as
revised by W. Creighton Schwan. — 39th ed.
      p. cm.
    Includes index.
    ISBN 0-9603294-7-1

    1. Electric wiring, Interior.   I. Schwan,
W. Creighton.   II. Title.

TK3271.R53   1999   621.319'24
                              QBI99-221

Executive editor: Margaret R. Wolfe
Content editor: Carol J. Frick

For their review of the manuscript and helpful suggestions, thanks to:

Gordon Patterson, Ph.D., Department of Vocational and Adult Education, Auburn University

H. Brooke Stauffer, Director of Codes and Standards, National Electrical Contractors Association

Joseph Tedesco, Director, National Technology Transfer, Inc. Code Schools

Contents

Tables

Preface to the 32nd edition

This book has been written for people who want to learn how to install electrical wiring so that the finished job will be both practical and safe. The installation will comply with the National Electrical Code®. Then the finished job will be acceptable to electrical inspectors, power suppliers, and others having jurisdiction in the matter.

Electrical wiring cannot be learned by skimming through this or any other book for 15 minutes; neither should the book be considered a "Quick Answer" reference. Nevertheless, careful study of this book should enable you to wire a house or a farm so that it will be acceptable to everyone concerned. However, before doing any wiring, learn how to do the job correctly.

The author hopes this book will also be of considerable value in planning a wiring job, to enable you to write sensible specifications that will lead to your securing maximum usefulness from electric power—now, and 10 years later. Careful planning will avoid later changes that usually cost several times as much as when included in the original plans.

Throughout this book I have emphasized the reasons why things are done in a particular way. This will help you to understand not only the exact problems discussed, but will also help you to solve other problems as they arise in actual wiring of all kinds.

The first edition of *Wiring Simplified* appeared in 1932. This 32nd edition (and all future editions) will be prepared by W. Creighton Schwan, who is especially well qualified to do so. I am very grateful to him for undertaking this project.

—H. P. Richter

Preface to the 39th edition

The ability of H. P. Richter to reduce a complex subject to simple, easy to understand terms is rare. Matching that ability is improbable, but it is a challenge I seriously accept. This 39th edition reflects revisions made necessary by the many changes in the 1999 National Electrical Code® and by materials and methods not covered in previous editions.

Comments and suggestions for improvements to the book are greatly appreciated.

—W. Creighton Schwan

Introduction

Wiring Simplified meets the needs of a variety of readers—

■ The homeowner doing a wiring installation in a new home or an update of existing wiring in an older home is given instructions each step of the way to completing a safe, practical job.

■ Someone deciding whether to tackle a job or hire a professional will learn what is involved in doing the job as well as what materials and tools are needed.

■ The person wanting to troubleshoot and make electrical repairs will find instructions in a special chapter.

■ Consumers of electric power will gain an understanding of wiring systems and the roles played by the user and the supplier in the handling and consumption of electricity.

■ Students and teachers in a variety of educational settings will find this a valuable text.

A new look and a major revision with 44 added pages mark this 39th edition of *Wiring Simplified*. Throughout its history of more than 65 years of continuous publication, the book has been updated to conform to each new National Electrical Code. This latest edition reflects the 1999 NEC.

Now all the information that longtime users have come to depend on in this authoritative book is easier to find and use. The depth of content and basic order of presentation developed by H. P. Richter are retained, with a few topics moved to new locations and many chapters renamed to better describe what is covered. Hundreds of new descriptive paragraph headings stand out in the text for quick reference.

Instructions on wiring procedures are enhanced by 225 drawing and photos, many of them new or revised. An expanded table of contents and index quickly lead the reader to specific topics, and the streamlined language is easy to understand. The new division of the book's 19 chapters into four parts take the reader from the basics to installing a wiring system and on to special projects.

Guide to using this book

Wiring Simplified is divided into four parts that cover the various phases of a wiring job—from understanding basic vocabulary and principles, to planning, installation, and finishing. Each part provides necessary foundation for the parts that follow.

> GETTING STARTED—THE BASICS
> WIRES, CIRCUITS, AND GROUNDING
> INSTALLING SERVICE EQUIPMENT AND WIRING
> SPECIAL WIRING SITUATIONS AND PROJECTS

The best way to use this book is to read it through from start to finish and then keep it in your toolbox or on a hook in your workshop as a convenient reference for specific procedures. The table of contents at the front of the book provides an overview of the wide range of material covered. This and the index at the back are your keys to locating topics quickly and easily. Citations of specific National Electrical Code (NEC) requirements appear under many index topics. Most citations reflect references appearing in the text of *Wiring Simplified*. In cases where there is a broader discussion in the NEC than is possible in *Wiring Simplified*, the citation is directly to the appropriate NEC material. The glossary at the back offers concise definitions to many commonly used wiring terms. Be sure to keep up with changes in the National Electrical Code by getting each new edition of *Wiring Simplified*, which is updated every three years to correspond to the new National Electrical Code.

GETTING STARTED—THE BASICS provides basic background necessary to understand the safety, practical, and legal considerations involved in wiring decisions and practices. This is essential reading before undertaking any electrical work. The standards and codes that help ensure safety are explained, and important safety precautions are offered. Practical goals to consider in the planning of any wiring job are described in plain language. A bare-bones outline of how electrical power is measured and delivered to the home provides basic vocabulary for understanding issues of usage, cost, and conservation of electrical power.

WIRES, CIRCUITS, AND GROUNDING explains how to choose the wire types and sizes, electrical devices, and the tools you need for a particular job. Proper overcurrent protection is given particular attention. The chapter on circuit diagrams shows a simple way to use wiring diagrams for designing an installation of any size. An entire chapter is devoted to the importance and specifics of grounding, and the emphasis on grounding is reinforced throughout the rest o f the book.

INSTALLING SERVICE EQUIPMENT AND WIRING puts to work the information covered in previous chapters on safety practices, equipment selection, and technique. This installation phase is when *Wiring Simplified*'s convenient size is a real plus— just put the book in your toolbox and take it right to where you're working for ready reference. The chapter on installing the service entrance explains the details involved in establishing the connection point between the power supplier's wires and your electrical installation. Step-by-step instructions are given for selecting and installing the boxes and switches needed in the circuits you have planned; for making the wiring connections at the boxes, switches, and receptacles; for running cable or conduit; and for testing the completed installation. You will also find information on how to wire detached garages and accessory buildings such as studios, greenhouses, or machine and woodworking shops. The challenges of modernizing an outdated wiring system are described, with many expert hints for making the task easier. Wiring requirements for a variety of specific household appliances are listed. The details of finishing an electrical installation wrap up this professional approach to getting the job done.

SPECIAL WIRING SITUATIONS AND PROJECTS begins with a discussion of the types and wiring requirements of stand-alone motors commonly used in residences and on farms. The chapter on farm wiring covers the unique considerations of rural electrical installations such as constructing the meter pole, wiring animal buildings, and rural safety issues. A new chapter on wiring for low-voltage systems such as telephones and doorbells will be of interest to many who might not be ready to attempt other types of wiring installations. The last chapter, on troubleshooting and repairs, is a guide to diagnosing and remedying common problems such as blown fuses or tripped breakers, broken switches and receptacles, and lamps and doorbells that don't work—information that can be used by people who never intend to undertake an electrical installation but want to be able to maintain the electrical equipment they depend on each day.

CHAPTER 1

Standards, Codes, and Safety

WE DEPEND ON ELECTRIC POWER so extensively in our homes and businesses, on our farms, and in industry that it would be difficult to list all the uses that we make of it. Electric power serves in countless ways with little danger to the user, so we often take it for granted. Yet when not used properly electric power can cause fire, destroy property, seriously injure people, and even cause death.

STANDARDS

Electric power is safe for us to use only when it is under control. We control it by using wiring and electrical equipment that is of dependable quality. Equally important, the wiring and electrical equipment must be properly installed and maintained.

Electrical parts and devices are manufactured according to specific safety standards. They should be installed using methods that are uniform throughout all the states—the methods proved by experience to be both practical and safe.

Underwriters Laboratories Inc.® (UL)   Reputable retailers and manufacturers sell only merchandise that is "Listed by Underwriters Laboratories Inc." or "Listed by UL" as it is commonly abbreviated. Underwriters Laboratories Inc. is an independent, not-for-profit testing laboratory. When a manufacturer submits its product to UL, the product is investigated and subjected to performance tests. If it meets the safety standards, it is then "Listed by UL." (Many people say "Approved by Underwriters" but that is not correct terminology.) To determine that the product continues to meet UL's requirements, trained field representatives periodically visit the factory where the product is made to audit production controls. Additionally, UL regularly tests factory and store samples. If these samples meet the requirements, the product continues to be Listed and will bear the UL Listing Mark.

Some items (wire, large switches, fixtures, conduit) have a UL Listing Mark similar to those shown in Fig. 1–1 on each coil or piece. Cord sets (extension cords) have labels as shown in Fig. 1–2. Devices like toggle switches carry the name of Underwriters Laboratories Inc. (may be abbreviated) and the word "LISTED" molded or stamped on each piece. Still other items such as receptacles, sockets, outlet boxes, and toasters have no label, but the nameplate, tag, or the item itself bears the Listing Mark of Fig. 1–3. Each piece is marked in some way so that the inspector or purchaser can identify the manufacturer by referring to the UL Directory, which lists manufacturers that have proved their product meets the UL requirements. The UL Listing Mark on the product indicates compliance.

Some inspection authorities accept third-party certification by other organizations such as Factory Mutual (FM), ETL Testing Laboratories Inc. (ETL), Canadian Standards Association (CSA), and some others, but for the electrical materials and equipment discussed in this book, UL is by far the most widely recognized.

ENCLOSED INDUSTRIAL CONTROL PANEL
No.

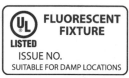

FLUORESCENT FIXTURE
LISTED
ISSUE NO.
SUITABLE FOR DAMP LOCATIONS

INSULATED WIRE
LISTED
ISSUE NO.
500 FT.
PRINTED IN U.S.A

ENCLOSED SWITCH
LISTED
No.

Fig. 1–1 Examples of labels applied on merchandise Listed by Underwriters Laboratories Inc.

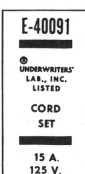

E-40091

UNDERWRITERS' LAB., INC. LISTED

CORD SET

15 A.
125 V.
1875 W.

E-40091

◄**Fig. 1–2** Listed cord sets (extension cords and appliance cords) are tagged with a flag label.

LISTED

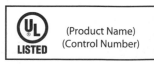

(Product Name)
(Control Number)
LISTED

Fig. 1–3 If because of size, shape, material, or surface texture the product bears only the round symbol at left, a complete Listing Mark with the four elements shown at right will be found on the smallest unit container in which the product is packaged.

"LISTED" usage requirements   "LISTED by UL" indicates the merchandise is suitable and safe *if used for the purpose and under the conditions* for which it was designed. Electrical inspectors will turn down Listed merchandise if improperly used. For example, the inspector will refuse to accept Listed armored cable used in a barn if the barn is a corrosive location, because armored cable is for use only in permanently dry locations. Listed lamp cord will be turned down if used for permanent wiring because such cord is Listed only for use on portable equipment.

Listing by UL does not mean that two similar pieces of merchandise are of the same quality. It merely indicates that both pieces meet the UL safety requirements. One may far surpass those requirements, while the other just barely meets them. For example, of two brands of toggle switches, one may average 25,000 ons and offs at full load before breaking down, and the other only 6,000. Use your own good judgment in making your choice among Listed brands just as you would in selecting other merchandise.

CODES, PERMITS, AND LICENSES

Listed electrical parts of high quality that are carelessly or improperly installed may still present a risk of both shock and fire, so rules have been developed regulating the installation of wiring and electrical devices.

National Electrical Code®   For a safe installation, Listed devices must be installed as required by the National Electrical Code (abbreviated NEC). The NEC is simply a set of rules specifying the installation and wiring methods that over a period of many years have been found to be safe and sensible. The NEC permits installations to be made in several different ways, but all wiring must be done in one of the ways described in the NEC. A revised edition of the National Electrical Code is published every three years, with the next edition to appear in 2002. Every reader is urged to obtain a copy of the NEC and to use it as a reference. It contains detailed regulations and extensive tables. Because the NEC is more than 600 pages long, only its most important general points can be included in *Wiring Simplified*. If you are using any of the conduit or tubing wiring methods, it is important to have your own copy of the NEC because the tables for the number of wires in conduit and tubing occupy 22 pages and cannot be included in *Wiring Simplified*.

In this book, all reference to NEC section numbers will be to sections in the 1999 NEC, which contains a large number of changes from the 1996 edition. Many of these changes consist of: (a) renumbering a section without a change in text, so that the only change is in the number of the section;

National Electrical Code® and NEC® are registered trademarks of the National Fire Protection Association, Inc., Quincy, Massachusetts. Every reader is urged to study the NEC. You can obtain a copy by sending $46.75 plus $4.85 shipping and handling to—

National Fire Protection Association Batterymarch Park, MA 02269-9101

or by calling 1-800-344-3555. It is also carried in many large bookstores. Ask for *NEC 1999*.

(b) renumbering a section with changes in the text, sometimes of minor but often of major importance; and (c) new section numbers covering material not in the previous edition. Therefore, a 1999 section number mentioned in this book may or may not be about the subject covered by the same section number in an earlier or later NEC. Vertical lines in the margins of some NEC pages indicate material that differs from the previous edition. A black dot appears in the margin where material has been deleted (but be careful—it may have been relocated to another place in the NEC and not really deleted at all).

All methods described in this book are in strict accordance with the latest National Electrical Code as interpreted by the author. Every effort has been devoted to making all statements in the book correct, but the final authority on the NEC is your local inspector. This book covers only the wiring of houses and farm buildings. Any statement that some particular thing is "always" required by the NEC means "always so far as the type of wiring described in this book is concerned."

Study of the NEC is necessary and helpful, but the NEC alone will not teach you how to wire buildings. Read NEC Section 90-1 for the exact wording of its purpose, which is essentially the practical safeguarding of persons and property from hazards arising from the use of electricity. These hazards are fire and electric shock. The NEC contains provisions considered necessary for safety. Compliance with the NEC together with proper maintenance will result in an installation essentially free from hazard but not necessarily efficient, convenient, or adequate for good service or future expansion. The NEC is not intended as a specification for design, nor is it intended as an instruction manual for the inexperienced.

Legal aspects   Neither the NEC nor the UL requirements have the force of law. However, states and cities pass laws requiring that only UL Listed materials may be used in wiring, and that all wiring must be in accordance with the National Electrical Code. Usually power suppliers will not furnish power to buildings that have not been properly wired, and insurance companies may refuse to issue policies on buildings not properly wired. Therefore you have no choice except to follow the law, and in doing so conscientiously you will automatically produce a safe installation.

Local codes   The National Electrical Code is sometimes supplemented by local codes or ordinances which are seldom contrary to the national code but limit its application. For example, armored cable wiring is one method permitted by the NEC but sometimes prohibited by local codes.

Local permits   In many places it is necessary to get a permit from city, county, or state authorities before a wiring job can be started. The fees charged for permits generally are used to pay the expenses of electrical inspectors, whose work leads to safe, properly installed jobs. Power suppliers usually will not furnish power until an inspection certificate has been turned in.

Local licenses   Many areas have laws that no one may engage in the business of electrical wiring without being licensed. Does that mean that you cannot do electrical work on your own premises without being licensed? In some localities, the law is interpreted that way, but not in others. Consult your power supplier or local authorities.

Remember that if a permit is required, you must get one before proceeding. Before applying for a permit, be sure you understand all problems in connection with your job so that your wiring will meet national and local code requirements. If it does not, the inspector must turn down your job until any errors have been corrected.

SAFETY PRACTICES

Electrical systems in buildings must be installed in such a way that they present the least possible hazard to the occupants or the property. Any unique conditions and uses, such as encountered on farms, must be carefully evaluated for wiring requirements that promote safety for people, animals, equipment, and buildings. In addition, the work itself must be performed safely for the protection of the installer. Unless you feel confident that you can satisfy these safety objectives, it would be best that you use this book as a guide to understanding your wiring system and leave the actual installation to a professional.

Protect your eyes   For your personal safety whenever you are doing any electrical work, protect your eyes at all times by wearing safety glasses. Bits of copper can fly out when cutting wire, and hot solder, flux, and plaster dust have a way of making right for the eyes. If an arc is inadvertently started when you must test on an energized circuit, molten metal can be thrown out too quickly for you to escape it.

Be aware of electrical fire hazards   Short circuits, overheating, and the arcing or flow of electricity where it doesn't belong—these are hazards that can result from faulty wiring and cause fire. Be careful to use wires with the proper ampacity (explained in Chapter 4) and check for wires or cords with damaged insulation. In Chapter 7 you will find a thorough discussion of the essentials of proper grounding. Protecting your electrical installation through the use of overcurrent devices— fuses and circuit breakers—is the focus of Chapter 5. Limiting vulnerability to lightning damage is a special concern for rural residents and is discussed in Chapter 17: *Farm Wiring*. Safe wiring practices for fire prevention are emphasized throughout the book.

Take advantage of hazard warning detectors   Faulty wiring is just one of many possible sources of fire or other hazards in a home. Installing appropriate alarm equipment is a relatively small investment that can save lives. Regularly check to see that your warning alarms are functioning. Get in the habit of replacing the batteries in a smoke detector or CO detector every fall at the same time as you reset your clocks to standard time. Remember that even permanently wired detectors depend on battery backup.

Smoke detectors   Self-contained smoke detectors that sound an alarm when visible or invisible combustion products are sensed have been credited with saving many lives by alerting occupants early enough for them to escape from a fire. Smoke detectors in homes are required by local building codes. There are three types of detectors. *Ionization alarms* are especially effective at detecting fast-flaming fires which quickly consume combustible materials such as burning paper or grease. *Photoelectric alarms* are more effective at detecting slow, smoldering fires such as a cigarette burning in a couch which may take hours before bursting into flame. For your maximum protection, install both types or select the third kind—a *dual detector* that combines both features. Building codes in each community specify the required locations but generally not the type of alarm. Some local codes require a smoke detector in the hallway outside bedrooms or above the stairway leading to bedrooms on an upper floor, while other localities require a detector in each bedroom, the kitchen, and one on each floor of the dwelling.

Battery-operated, plug-in, and direct-wired smoke detectors are available. Battery-powered models may be acceptable in an existing dwelling, but in new construction most building codes require connection to the electrical source, with a battery backup and all alarm units interconnected by wires so that if one sounds off, they all sound off. The required location is on the ceiling, or on the wall not more than 12 inches from the ceiling. Don't place the detector in the path of ventilation which would move air past the detector faster than in other parts of the room. To install a direct-wired detector, choose a circuit with often-used lights on it, such as the bathroom light, to assure the circuit could not be off without being noticed. Wire the detector directly across the two circuit wires, unswitched.

Carbon monoxide (CO) detectors   Many local ordinances also require permanently wired carbon monoxide detectors with battery backup to alert occupants to the presence of this invisible, odorless, deadly gas. Check with your building official or fire marshal for specific requirements. CO detectors are highly recommended even if not required by local codes. Battery-only styles are available. CO detectors can usually be found in stores near the smoke detectors.

Arc-fault circuit-interrupter circuit breakers   The U.S. Consumer Product Safety Commission estimates that annually there are more than 40,000 fires in residential occupancies, resulting in 250 lives lost and $1 billion in financial loss. It has been estimated that 40 percent of these fires are caused by arcing faults—the unwanted flow of electricity through an insulating medium (such as air). An arc can generate enough heat to be a source of ignition, but not enough current to trip the standard circuit breaker.

Just as there are circuit breakers incorporating ground-fault protection for the purpose of preventing people from receiving dangerous shocks (see page 69), there are now circuit breakers incorporating arc-fault circuit interruption in addition to the normal overcurrent functions, and some that incorporate both GFCI and AFCI in the same breaker frame. Arcing faults exhibit a characteristic current and voltage pattern that can be detected by the electronics in the AFCI which then interrupts the circuit. In order to allow time for field experience with this new product, even though they are UL Listed, the effective date in the National Electrical Code for use on circuits supplying receptacle outlets in dwelling unit bedrooms is January 1, 2002, but there is no prohibition to using them now on any receptacle outlet circuit.

Prevent electric shock   Safe installation practices are emphasized throughout this book. Shock—the accidental flow of electric current through the body—is the principal hazard that can result from unsafe practices. Observing the cautions that follow will help you avoid injury.

Disconnect the circuit   Whenever possible, work should be done only on circuits that have been de-energized (disconnected from the source of power). If the disconnecting means is out of sight from the work location, you must take positive steps such as taping the switch to the OFF position or removing the fuse to make sure that someone else cannot inadvertently energize the circuit while you are working on it. Some testing can be done only on energized circuits. Be extremely careful when testing that you do not contact live parts. As explained on page 60, receiving a shock is more dangerous when the victim is standing in water or a damp location than on a dry surface. For your protection, stand on dry boards or a rubber mat and be sure that neither you nor the floor is wet. Remember that electricity and water are a dangerous combination.

Avoid shocks from small appliances   Even when a building has been wired in strict accordance with the NEC, shocks can still occur from the use or misuse of appliances. Fatal shocks have occurred in bathtubs or showers when an appliance such as a radio or hair dryer has accidentally fallen into the water. In that situation, even an appliance that is not defective can cause a fatality. Touching a defective radio or heater or other appliance while in a tub or shower can also be fatal. A child might chew on the cord of a lamp or appliance; in doing so, if the child punctures the insulation of the cord (or if the cord is defective) while he or she is also touching a grounded object such as a radiator or plumbing, a fatal shock can result. A person using a defective appliance while touching a grounded part of plumbing can easily receive a shock.

A toaster can present extreme danger if the toast sticks rather than popping up when it is done because when that happens people sometimes use forks to release their toast. Don't do it without first unplugging the toaster. The live heating elements are exposed in the toaster. If your fork touches a heating element and you are touching a grounded object (such as a water faucet or a sink) at the same time, you could receive a dangerous or even fatal shock. Unplug the toaster before reaching inside it for any purpose, or use wooden tongs to extract the toast.

Know how to help shock victims safely   If you discover a shock victim, the most important point is to be careful not to become a second victim yourself. This book cannot possibly give you all the needed information, but it can give you a few hints.

We'll first discuss shocks caused by the ordinary 120/240-volt wiring in a house or a farm building. If you find a victim and it appears that the shock was caused by an appliance or similar equipment, disconnect that equipment. If it has a wall switch, turn it off. If it doesn't, pull the plug out of its receptacle; do not touch the cord because it may be defective, but rather grasp the plug and pull it out. Do not touch the appliance itself.

In cases where the source of the shock is not apparent, hurry and open the main switch of the building, or throw the main circuit breaker to OFF if your building is equipped with breakers. Then turn your attention to the victim. If the victim is unconscious or appears not to be breathing, call "911" or the emergency service in your area for assistance. If you know CPR (cardiopulmonary resuscitation), start using it while you are waiting for medical help. (You'll be better prepared for emergencies if you have taken a class in CPR. Contact your local Red Cross chapter for information.)

If the shock has been caused by a high-voltage line that has been knocked down by a storm or accident, the situation is quite different and many times more dangerous. In such a case, phone your power company. If the high voltage wire is touching the victim, *do not* touch him or her; if you do, you could become the second victim by trying to help somebody who quite possibly has received a fatal shock. Theoretically you can use a completely dry wooden pole at least four feet long to push the fallen wire off the victim, but since such accidents seldom occur near your home, you will not have such equipment with you. Depend on experts from your power company to perform the rescue.

CHAPTER 2

Planning Your Electrical Installation

THE PREVIOUS CHAPTER on standards and codes discussed the importance of following the National Electrical Code (NEC) strictly in order to produce a safe wiring installation. This chapter outlines how careful planning will help you make it a convenient, efficient, and practical installation. Here you will find the characteristics of an adequate house wiring installation, the principles of good lighting, and NEC requirements for locating lights, switches, and receptacles.

CONSIDER PRESENT AND FUTURE NEEDS

Plan your electrical installation so you will still be pleased with it in the years ahead. Do not skimp on the original installation. Adding outlets, receptacles, switches, fixtures, and circuits later usually costs several times more than it would to include them in the original job.

Look ahead to the equipment you are going to be using five or even fifteen years from now. Besides the basics of electric cooking, water heating, and air conditioning, consider items such as computers, entertainment centers, workshop and home business equipment, and even appliances not yet on the market. By installing large wires and extra circuits now you will have adequate wiring in the future. You will be making a good investment.

Install large service entrance   All the power you use comes into the building through the service entrance wires and related equipment. Start your planning with a service entrance that will adequately handle present and future needs. Keep in mind that small service wires will not carry a large load satisfactorily. Chapter 8 contains a full discussion of the service entrance installation.

Benefit from larger circuit wires   For house wiring, the minimum circuit wire size permitted by the NEC is No. 14, protected by a 15-amp fuse or circuit breaker, except for door chimes and other low-voltage wiring described in Chapter 18. There is a trend toward using No. 12 wire, with 20-amp protection, and in a few localities it is required as a minimum. The larger wire means brighter lights, less power wasted in heating of wires, and less frequent blowing of fuses or tripping of breakers. The use of No. 12 wire in place of No. 14 adds very little to the cost of the installation and will prove a good investment. NEC requirements for various types of circuits are discussed on pages 44–47. Special circuits requiring No. 12 wire are explained under "Special small appliance circuits" on page 46.

LOCATION OF LIGHTING, SWITCHES, AND OUTLET RECEPTACLES

An adequately wired home is well-lighted. The lighting will be from floor and table lamps in some rooms and from permanently installed fixtures in others. Lighting fixtures do not need to be expensive or elaborate, but you must have outlets for fixtures located so they provide sufficient light where needed. In addition, plenty of switches and receptacle outlets are essential for convenience and safety.

Height from floor for switches and receptacles   There is no NEC requirement for the height from floor of switches and receptacles. The typical range is 48 to 52 inches for switches and 12 to 18 inches for general-use receptacles. (People with physical disabilities may require switches to be lower and receptacles higher than usual.) Kitchen receptacles are located in the backsplash above the counters; in the laundry room and the workshop a convenient height for plugging in washer, iron, and tools is 36 to 48 inches above the floor. In new construction make the heights uniform throughout. In a remodel or addition match the existing heights.

NEC requirements for lighting and switches   NEC Section 210-70(a) requires these areas to have some lighting controlled by a wall switch: every habitable room of a house; bathrooms; hallways; stairways; attached garages, and detached garages if wired; attics, underfloor spaces, utility rooms, and basements used for storage or containing air-conditioning, heating, or other equipment requiring servicing. In kitchens and bathrooms the switch must control permanently installed lighting fixtures. In other rooms the switch may control one or more receptacle outlets into which floor or table lamps can be plugged. Each entrance into the house must have an outdoor light controlled by a switch inside the house.

Regardless of NEC requirements, an adequately wired house will have wall switches located so you can come in through any entrance and move throughout the house from floor to floor without ever being in darkness or leaving lights turned on behind you. A wall switch (shown in Fig. 2–1) costs more than a pull chain, but the wall switch is far more convenient to use, and it provides accident insurance against stumbling while looking for a pull chain in the dark. Installing a pair of three-way switches so a light can be turned on or off from two different places (the bottom and top of a staircase for

Fig. 2–1 Typical toggle switch. The "plaster ears" on the ends of the strap are a great convenience in mounting the switch. *(Leviton Manufacturing Co., Inc.)*

example) costs only a few dollars more than a single switch and is well worth the difference in terms of safety and convenience. Providing convenient switches for lighting also helps to conserve energy since we are less likely to leave lights on if it is easy to turn them off when they are not needed.

NEC requirements for receptacles   NEC section 210-52(a) requires receptacle outlets in each kitchen, family room, dining room, living room, parlor, library, den, sun room, bedroom, recreation room, or similar rooms of dwelling units to be installed so that no point along the floor line in any wall space is more than 6 feet measured horizontally from an outlet in that space, including any wall space 2 feet or more in width and the wall space occupied by fixed glass panels in exterior walls. The wall space provided by fixed room dividers, such as free-standing counters or railings, is to be included in the 6-foot measurement. Hallways 10 feet or more in length must have at least one receptacle outlet.

Basements are required to have at least one receptacle outlet in addition to any outlets for laundry equipment. Attached garages—and detached garages having electric power—are also required to have at least one receptacle outlet. Every house is required by NEC Section 210-52(e) to have at least two outdoor receptacles—one on the front of the dwelling and one on the back—for yard tools, appliances, holiday lighting, etc. They must be no more than 6 feet 6 inches above grade level. See page 69 for locations where GFCI protection for personnel is required.

Fig. 2–2 Typical duplex receptacle, rated at 15 amps. It has two terminal screws on each side, plus one green grounding screw.

An adequately wired house has enough receptacles (shown in Fig. 2–2) so that you will never need an extension cord for a lamp, clock, or similar equipment. Extension cords are useful as temporary devices, but they should never be used to permanently carry power from a receptacle to lamps or other equipment. If there are circumstances where you cannot avoid using an extension cord, don't invite a fire by running the cord under a rug or carpet. Never run one across open floor space—stepping on it could result in damage and fire, and tripping on it could lead to injury. Never tack or staple a cord to a wall. It is best to reserve extension cords only for temporary use.

Fixtures and receptacles room by room   The following list details both the required and the recommended lighting, switches, and outlets to be installed in each room plus the garage and any accessory buildings.

Living room   Suit yourself about whether to have a fixture in the center of the ceiling or on a wall. Some homes depend entirely on floor and table lamps for living room light. If this is the case, it will be convenient to have one or two wall switches control all the receptacles into which the lamps are plugged so they don't have to be turned on and off individually. You might choose to have just the bottom or top halves of several duplex receptacles in a room controlled by a switch (see page 108).

No matter how small the room, there should be a minimum of five receptacle outlets for uses such as TV, stereo, radio, and lamps. Locate one receptacle where it will always be completely accessible for the vacuum cleaner regardless of the furniture arrangement. Many living rooms will be large enough to require more than five receptacles based on the "6 foot" rule. Regardless of room size, consider the probable arrangement of the furniture and place receptacles so they will not be difficult to reach. This may require an extra receptacle or two, but will be worth it.

Dining room   One ceiling outlet for a fixture should be provided; it must be controlled by a wall switch. Locate this fixture above the center of the dining room table rather than the center of the room. The "6 foot" rule usually will require at least four receptacle outlets. Be careful to locate them so they will not be hidden behind furniture. One receptacle should be always accessible from the dining room table for use with items like electric coffee makers, electric woks, and electric knives.

Kitchen   Abundant wiring is required in the kitchen. First, it should be well-lighted because it is often the center of activity. A ceiling fixture controlled by a wall switch must be provided for general lighting. If there are two entrances to the kitchen, use a pair of three-way switches so the light can be controlled from either door. Another light above the sink is necessary so a person does not have to work in his or her own shadow. Consider other kitchen work areas that may need specific lighting.

If a fixture is installed where it is possible to touch both the fixture and a water faucet or other grounded surface at the same time, the fixture should be made of porcelain or, if made of metal, it must be grounded. In such locations, fixtures should be controlled by a wall switch. These are safety measures. An ungrounded metal fixture or metal pull chain could become energized through a fault in the fixture and make a serious shock possible.

The NEC requires two special circuits to handle only receptacle outlets for small appliances (including refrigeration equipment) in kitchen, pantry, dining room, and breakfast room. Both of these required small-appliance branch circuits must appear in the kitchen. Install one outlet for the refrigerator, and place the others 6 to 10 inches above the counter in the backsplash. A counter space 12 inches wide requires one receptacle, and others are required so that no point along the counter space is more than 24 inches from a receptacle outlet. Spaces separated by a sink or range top are considered as separate spaces. Because the lack of a backsplash makes receptacle placement at counter peninsulas and islands difficult, the spacing requirement is relaxed to require only one receptacle outlet per counter. Be aware that where receptacles are located in the face of the cabinet within 12 inches below the counter top, cords will be draped over the counter edge providing children with a ready means to pull over an appliance that might cause injury to them. See page 69 for locations where GFCI protection is required.

Install a range receptacle (discussed on pages 145–147) in the beginning, because installing it later will be much more costly and difficult.

Bedrooms   Some people want a ceiling light in each bedroom; others do not. If you install a ceiling light, have it controlled by a wall switch near the door. If there will be no ceiling light, be sure some of the receptacles are controlled by a wall switch. The "6 foot" rule will probably require at least four receptacles. Plan location carefully; one should be permanently accessible for the vacuum cleaner, and another near the bed for items such as reading lamp, clock radio, and heating pad.

Closets   Don't overlook closet lights. Install the fixture on the ceiling or on the wall above the door so it is separated from the storage space by 12 inches, defined as at least 12 inches out from the walls, or the actual width of a shelf if greater than 12 inches, all the way to the ceiling. This is a safety measure. Many fires have been traced to clothing touching bare lamps in closets. The temperature of the glass bulb of an incandescent lamp is often higher than 400° F. In no case can a pendant fixture or an incandescent lamp without an enclosure be installed in a closet. If the fixture is a recessed incandescent with a solid lens, or a surface-mounted or recessed fluorescent fixture, the distance from the storage space may be reduced to 6 inches.

Bathrooms   A ceiling light controlled by a wall switch is essential. One additional light above the mirror at the basin is often installed, but that is not good lighting. Two lights—one on each side of the mirror—are needed for applying makeup or shaving. Fluorescent brackets are ideal for the purpose. Pendant fixtures, ceiling paddle fans, and track lighting are prohibited in the area within 3 feet horizontally and 8 feet vertically of the tub or the shower area. A GFCI-protected receptacle is required within 36 inches of the outside edge of each basin (see page 69). A dedicated 20-amp circuit is required for all the bathroom receptacles.

Do not locate receptacles near the bathtub. Each year people are killed in the tub when an appliance falls into the water. The appliance may not be defective, but it is just as dangerous as a defective one would be. A built-in electric heater is safer and more convenient than a portable one.

Stairs   Where there are stairs, locate a fixture so it will light every step. Use three-way switches so the light can be turned on and off from both the top and bottom of the stairs. This is inexpensive insurance against accidents.

Hallways, entrances, and basement   Every hallway deserves a light. For a long hallway, install three-way switches to control the light from either end. Install a receptacle for a vacuum cleaner. At each entrance to the house, outdoor lights controlled by switches indoors are essential. A motion-sensitive light is an option for your outdoor lighting. When people approach the door from outside at night, the light will turn itself on automatically. This is especially useful if the house does not have an attached garage. In the basement, install numerous receptacles for washer and dryer, workshop motors, and similar uses.

Garage and accessory buildings   If the garage is attached to the house, treat it as another room. Three-way switches are preferred for convenience, one at the garage door and the other at the door between garage and house. There are several considerations for a garage that is separate from the house. Most of these considerations apply to other accessory buildings as well.

The minimum wiring in a garage is a light turned on or off by only a switch in the garage and fed by a pair of wires from the house. Most people will want a light that can be controlled from both the garage and the house, requiring a three-way switch at each end.

NEC Section 210-52(g) requires every garage to have at least one GFCI-protected receptacle outlet. More than one may be desirable for a battery charger, tools, etc. In garages and accessory buildings, receptacle outlets that are not required to be GFCI-protected include outlets that are not accessible, such as a ceiling outlet for a garage door opener, and outlets for appliances in dedicated space, such as a freezer. All others must be GFCI-protected. In a garage, it is very convenient to have receptacle outlets that are always live rather than receptacles that are turned off with the garage light. With receptacles permanently on, you can plug in a battery charger and have it work all night without having the garage light on. See pages 110–112 for procedures and wiring diagrams for detached buildings.

CHOOSING LIGHTING FIXTURES

Good lighting requires not only sufficient light but also proper distribution of light throughout the room. It is easier to read in the shade of a tree than in direct sunlight even if there is less light in the shade. In direct sunlight, the light all comes from one point—the sun. In the shade of a tree, it comes from all directions. Similarly, fixtures with exposed lamps produce light from one point, causing sharp shadows and glare that make reading or sewing or other close work difficult. Provide well-diffused "shade of the tree" light by using fixtures that do not have exposed lamps.

For best lighting, ceilings should be a light color. A white ceiling reflects most of the light thrown against it. Ivory is not quite as good for the purpose. A flat finish is better than a glossy finish.

Some fixtures provide good illumination and others are primarily decorative. If you have good floor and table lamps to provide plenty of lighting where people sit, there is no reason why fixtures in the same room cannot be the decorative kind that provide some general lighting but not enough for exacting work. When ceiling fixtures are to be the primary source of light in a room, take care to locate them where they will provide lighting for critical seeing tasks. Buy the styles of fixtures that suit your taste and budget. Make sure the fixtures you select are electrically good and safe—look for the UL Listing label on each fixture. Good lighting is not necessarily expensive. Lighting fixtures are available in a wide range of prices, and in many places a 100-watt lamp can be burned for a penny an hour.

Lamps (light bulbs)   Bulbs? Lamps? Different people call them by different names. Lamps are commonly bought and sold as "light bulbs" but technically only the outer glass globe is a bulb, while the entire unit is a lamp. In this book, they are called lamps in accordance with the terminology used in the NEC. See Fig. 6–1 for the symbol used to indicate a lamp in diagrams used in this book.

Types and efficiency   There are many types of lamps: incandescent, fluorescent, metal halide, mercury vapor, sodium vapor, etc. Three 60-watt lamps (total 180 watts) give 10 percent more light than five 40-watt lamps (total 200 watts). One 150-watt lamp gives twice as much light as five 25-watt lamps (125 watts). One 100-watt lamp gives 15 percent more light than three 40-watt lamps (120 watts). In other words, a fixture that uses one large lamp will in general provide more light than one using several small lamps of equivalent total wattage.

Labeling   Federal regulations require most of the commonly used incandescent and fluorescent lamps to be labeled with the light output measured in lumens, energy used measured in watts, and the life in hours. To save energy, choose a lamp based on the light output and then select the lowest wattage lamp that will deliver that light.

Incandescent lighting   Ordinary lamps have an average life of 750 to 1,000 hours if used on circuits of the voltage stamped on the lamps, which is the voltage for which they were designed. There are also several newer types on the market.

"Long life" incandescent lamps   It is a simple matter to increase the life of a lamp by burning it at a voltage lower than it was designed for, as is done in "long life" lamps. For example, a lamp designed for 135 volts but burned at 120 volts will last about four times longer; a 140-volt lamp burned at 120 volts will last about eight times longer. But there is a catch—the light per watt is reduced by 20 to 30 percent and the cost is increased by 25 to 35 percent. In addition, such "long life" lamps sometimes consume more watts than the number stamped on them. If such a lamp is marked "60 watts" but actually consumes 70 watts, it will seem to produce just as much light as a standard 60-watt lamp with a 1,000-hour life. There is a place for such "extended service" lamps if they are located where they are difficult to replace. Some lamps have reduced wattages. For example, replacing a 75-watt lamp with a 67-watt lamp will certainly save energy—but it will also give less light.

Energy-efficient incandescent lamps   There have been some real advances in incandescent lamp efficiency. Bulbs filled with krypton gas instead of nitrogen reduce energy consumption by 10 to 20 percent with no noticeable decrease in lamp life or light output. The lamp-within-a-lamp has a tungsten-halogen lamp inside a compact quartz tube within a conventional lamp bulb. The halogen gas redeposits tungsten particles back onto the filament, reducing blacking and providing up to three times the life of a conventional incandescent lamp.

Although the initial cost is higher, these energy-efficient lamps provide a return on investment of up to five times over their life. They last longer and use less electricity, saving you money and conserving energy at the same time. They come in 42-watt,

52-watt, and 72-watt sizes to replace standard 60-watt, 75-watt, and 100-watt incandescents. Their average life is 3,500 hours.

Fluorescent lighting   Fluorescent lighting has many advantages over incandescent. It produces far more light per watt of power used, and fluorescent lamps last much longer than ordinary lamps. The life of fluorescent lamps depends mostly on the number of burning hours per start. If turned on and left burning continuously, they will last at least eight to fifteen times as long as ordinary lamps. Even if turned on and off frequently, the fluorescent will last three to five times as long as an ordinary lamp. Fluorescent lamps used in homes come in lengths from 18 to 60 inches, consuming from 15 to 60 watts. The 48-inch/40-watt is the most common.

Brightness and whiteness   A further advantage of fluorescent lamps is that their surface brightness is much lower than that of filament lamps, providing light that is more diffused and thus more comfortable. Standard fluorescent lamps come in an assortment of different kinds of "white," including deluxe warm white, warm white, white, cool white, and deluxe cool white. The cool white is most commonly used. It produces light most like natural light and emphasizes the blue and green colors in objects lighted. The warm white produces light more like that of ordinary filament lamps and emphasizes red and brown. The deluxe varieties emphasize these qualities still more but are a bit less efficient, producing a little less light per watt consumed.

Energy-efficient fluorescent lamps   There are energy-efficient lamps and ballasts, including solid-state ballasts, that have ratings other than "standard," and these are truly more efficient than the lamps and ballasts they replace. Compact fluorescent lamps as small as 7 watts can replace incandescent lamps in many applications at great savings in energy. These compact lamps are made with a self-contained ballast and an Edison screw-shell base, so they can very simply replace an incandescent lamp. A compact fluorescent rated around 25 watts replacing a 100-watt incandescent ranges in price from $10 to more than $20. A 15-watt compact fluorescent lamp replaces a 60-watt incandescent, a 20-watt fluorescent replaces a 75-watt incandescent, and a 25-watt fluorescent has light output equivalent to a 100-watt incandescent. These fluorescent lamps have an average life of 10,000 hours. Another type is the two-piece circular compact fluorescent lamp that has a reusable electronic ballast. It comes in 22-watt and 30-watt sizes to replace 100-watt and 150-watt incandescent lamps.

Compact fluorescent lamps come in many different styles and shapes. Newer types are labeled "electronic" because of the type of ballast used. Some have large bases that may not fit within the harps of your table and floor lamps, so measure the harp width before your trip to the hardware or home center store where the compact fluorescents are available. (Some electric power suppliers also sell them.) Generally a square style will fit in a smaller space than a circular shape. The lamps cannot be used with dimmers, electronic timers or photocell devices. Read the package before purchase to make sure you can use the lamp for your intended purpose.

Operation   In fluorescent lamps, the electric current does not flow through a filament as in ordinary incandescent lamps, but rather jumps as an arc from a contact in one end of the lamp, through a gas in the glass tube, to another contact at the other end of the lamp. If you had such a fluorescent lamp made of clear glass in a dark room, you would see only a slight blue glow when the lamp was turned on. The arc in the lamp provides mostly ultraviolet light which is invisible. The inside of the lamp is coated with a special powder that fluoresces (glows) when struck by ultraviolet light, producing the visible light we see.

The electrical operation of a fluorescent lamp is a bit complicated. The complete circuit includes the lamp, a ballast, and a starter. The most common fluorescent lamp has a filament ("cathode") at each end, but these are lighted only for a second or so when the light is first turned on. The ballast—an electrical winding on a steel core—does two things: it limits the total power that can flow through it, and it also has the peculiarity that when power flowing through it is disconnected, it momentarily delivers a voltage much higher than the original voltage flowing through it—for a brief moment it becomes a step-up transformer. The starter is a switch that opens itself after power has flowed through it for a moment, and then remains open. (The magnetic ballast described is gradually being replaced by an electronic ballast, which accomplishes the same ends by different and quieter means.)

Fig. 2–3 shows the circuit for the first second or so after the light is turned on. The current flows first through the ballast, then through the filament at one end of the lamp, then back to SOURCE. The starter opens itself, and after that there is no circuit other than the lamp that the current can flow through. As the switch in the starter opens, the ballast delivers a momentary high-voltage kick that is enough to jump the gap inside the lamp; the current then flows through the lamp from end to end. See Fig. 2–4 showing the circuit after the switch in the starter has opened and the lamp is in normal operation. A bit of mercury in the tube vaporizes when the lamp goes into operation, helping the current flow through the lamp. The ballast limits the current through the lamp, keeping it at a proper value. Some fluorescent lamps, such as the rapid-start and instant-start types, start somewhat differently. The basic principle of operation is the same in all varieties—a high-voltage kick to start the lamp, with a ballast to limit the current to the proper value. The diagrams of Figs. 2–3 and 2–4 are for a fixture with only one lamp. Diagrams for fixtures with two or more lamps are more complicated.

Fig. 2–3 The current has just been turned on. The starter is still closed.

Fig. 2–4 The starter has opened, current flows through the tube; the lamp is lighted.

●

▶ **GETTING STARTED—THE BASICS**
WIRES, CIRCUITS, AND GROUNDING
INSTALLING SERVICE EQUIPMENT AND WIRING
SPECIAL WIRING SITUATIONS AND PROJECTS

CHAPTER 3

Measuring Electricity

WATER IS MEASURED in gallons, wheat in bushels, meat in pounds. Electricity cannot be poured into a measure or weighed on scales, but rather is something that must be considered as always in motion.

UNITS OF MEASUREMENT

We need to measure how much electricity flows past a given point at a given moment or in total over a period of time. To arrive at a measurement of electricity, the rate at which a quantity of it flows (coulombs per second, or amperes) and the pressure it is under when it flows (volts) are combined to arrive at wattage.

Amperes   The absolute measure of a quantity of electricity is the "coulomb." We can speak of electricity in terms of "coulombs per second" in the way we might speak of water in motion in terms of gallons per second. Few people may recall ever hearing the word "coulomb" because instead of "coulombs per second" we use the simpler term "ampere" (abbreviated "amp"). One ampere of electricity is defined as current flowing at the rate of one coulomb per second. (Note that we don't say "amperes per second," but just "amperes.")

An electric current is the flow of electrons past a given point. One ampere is the movement of 6.28 billion billion (6,280,000,000,000,000,000) electrons per second.

Volts   Water and air and other substances can be put under pressure which we commonly speak of in terms of pounds per square inch. Electric power is under pressure which is measured in volts (abbreviated "V"). Any ordinary dry cell or flashlight cell, when new, develops a pressure of about 1½ volts. One cell of a car battery develops 2 volts; six cells together develop 12 volts. Most house and farm wiring is at 120 volts for lighting and 240 volts for permanently installed appliances and for motors that run machinery. The voltage at which power is transmitted over high-voltage lines varies from 2,400 volts for short distances to 500,000 or more volts for long distances.

Watts and kilowatts   The total amount of power in a circuit at a given moment is expressed in "watts." Amperes alone or volts alone do not tell us the actual amount of power in a circuit. Both must be considered. Amperes and volts together tell us how much power is in a circuit at a given moment: volts × amperes = watts.

This formula is always correct with direct current, but it is correct only part of the time with alternating current. It is correct with lamps, ranges, toasters, and similar heating appliances. It is not correct with motors, or loads with transformers (radio, TV) or with ballasts (fluorescent lamps)—in all these the watts are somewhat less than volts × amperes. For ac load calculations, the NEC uses volts × amperes (VA) instead of watts.

Any wattage may consist of either a low voltage and high amperage, or a higher voltage and lower amperage. A lamp drawing 5 amps from a 12-volt battery consumes 60 watts (5 × 12 = 60); another lamp drawing ½ amp from a 120-volt line also consumes 60 watts (½ × 120 = 60 watts). The voltage and amperage differ widely, but the actual watts or power consumed by the two lamps is the same.

Watts measure power just as horsepower does. As a matter of fact, 746 watts is equal to 1 hp. A motor that delivers 1 hp delivers 746 watts and could just as well be called a 746-watt motor. (It uses more than 746 watts because some power is wasted as heat and it also takes some power to run the motor even when it is not delivering power). A lamp that uses 746 watts could just as well be called a 1-hp lamp.

A watt is a very small amount of power. When speaking of large amounts of power, it is simpler to speak of kilowatts (the Greek word "kilo" means thousand). One kilowatt (abbreviated "kW") is 1,000 watts. We speak of watts, not watts per hour or kilowatts per hour, just as we say that an automobile engine delivers 200 hp, not 200 hp per hour.

Watthours and kilowatthours   Watts and kilowatts measure the rate at which power is being used at any given moment. Watthours and kilowatthours, units of energy, measure the total amount of power that has been used during any specified interval of time. One watt used for 1 hour is 1 watthour. Multiplying the watts used by the number of hours gives the watthours. A 60-watt lamp used for 6 hours consumes 360 watthours (60 × 6 = 360). A 2,000-watt room air conditioner used for 2 hours consumes 4,000 watthours (2,000 × 2 = 4,000).

A watthour is a very small amount of energy, so it is common to speak of kilowatthours. A kilowatthour is 1,000 watthours. The air conditioner mentioned in the previous paragraph, consuming 4,000 watthours in 2 hours, consumes 4 kilowatthours (abbreviated "kW"). Electric energy from your power supplier is measured and paid for by the kilowatthour.

One kilowatthour will operate the average clothes washer for about 3 hours. It will operate a 1-hp motor for about an hour or pump about 1,000 gallons of water. It will operate the average stereo about 15 hours, a 50-watt lamp for about 20 hours, or an electric clock for about 3 weeks. One kilowatthour costs from 6 to 12 cents depending on your location.

RESIDENTIAL ELECTRICAL POWER

The amount of electricity that you and other consumers of electric power use leads to the cost you pay for electric service from your power company. Average 1998 rates for electric power are given on the next page. You can check your consumption on your electric meter (installed by your power company) and on your monthly statements. Methods to reduce costs include using energy-saving lamps (discussed on pages 15–16) and energy-efficient appliances (discussed on page 145). Examples of power consumption by various household appliances are given in Table 3–1.

Reading your meter Some meters being installed today provide a digital read-out of the total kilowatthours used by the household. You read a digital meter just as you do the odometer on your car.

Most meters have dials as shown in Fig 3–1. Two of the pointers move in one direction, the others in the opposite direction. Four dials are shown, but you may have five dials on your meter. To read the dial meter: Assume it is the beginning of the month; going from left to right simply write down the number that each pointer has passed, as on a clock. The total of the meter in Fig. 3–1 is 1,642 kWh.

Fig. 3–2 shows the same meter a month later. One of the pointers points directly to the 7. Before writing down 7, look at the pointer on the dial to the right; it has not quite reached the zero. Therefore, even though the pointer seems to point directly to 7, it has not actually reached the 7, so write down 6 instead, for a total reading of 2,269 kWh. (If the last pointer were just past the zero, the total would be 2,270.) The difference between the two readings, 627 in this case, represents the number of kilowatthours of electricity used during the month.

Fig. 3–1 To read this meter, write down the numbers the pointers have passed.

Fig. 3–2 The meter shown in Fig. 3–1, but now a month later.

GETTING STARTED—THE BASICS

Power rates   The rate charged for domestic electric power averages about 8.5¢ per kilowatthour in the United States, but the rate varies greatly. Before the energy crisis of the 1970s, rate schedules were stepped so the more power you used, the lower the average cost per kWh. Now, to encourage conservation of energy most rates are inverted—the more power you use, the higher the average cost per kWh.

Here is an example of a typical rate structure:

First 100 kWh used per month .... 4.8¢ per kWh
All over 100 kWh used per month .... 9.2¢ per kWh

The meter reading on the previous page showed 627 kWh consumed during the month. The bill is figured this way:

100 kWh at 4.8¢ $4.80
27 kWh at 9.2¢ $48.49
627 kWh .... Total $53.29
Average per kWh .... 8.5¢

Considering the rate of inflation and fluctuating fuel costs, these rates may not be typical at the time you are reading this.

Operating cost per hour   To find out how much it costs to operate any electrical load for one hour, multiply the watts that the item consumes by the rate in cents per kWh, placing the decimal point five places from the right to arrive at the cost in dollars per hour.

Some examples of operating costs:

ELECTRICAL LOAD	WATTS X RATE	OPERATING COST IN DOLLARS PER HOUR
60-watt lamp at 8.5¢ per kWh	60 x 8.5 = 510	$0.0051 (slightly more than a half-cent per hour)
600-watt appliance at 8.5¢ per kWh	600 x 8.5 = 5100	$0.051 (5.1¢ per hour)

How long does it take a particular electrical load to consume one kWh? To find out, simply divide 1,000 by the wattage of the load to get the number of hours, as in the following examples.

ELECTRICAL LOAD	1000 ÷ WATTS	TIME IT TAKES TO CONSUME ONE KILOWATTHOUR
40-watt lamp	1000 ÷ 40 = 25	25 hours
600-watt appliance	1000 ÷ 600 = 1.666	1.666 hours (about 1 hr 40 min)
electric clock using about 2 watts	1000 ÷ 2 = 500	500 hours

Watts consumed The following table will aid you in estimating power required (watts consumed), or the operating cost, for various appliances. The figures are only approximate, and new energy-saving appliances are appearing all the time. Appliances often have their total wattage (representing the load on the circuit) listed on their nameplates.

Table 3–1 WATTS CONSUMED BY VARIOUS APPLIANCES

FOOD PREPARATION	Watts
Blender	500 to 1,000
Coffeemaker	500 to 1,200
Convection oven	1,500 to 1,700
Dishwasher	1,000 to 1,500
Food processor	500 to 750
Frying pan	1,000 to 1,200
Hot plate, per burner	600 to 1,000
Knife	100
Microwave oven	1,000 to 1,500
Mixer	120 to 250
Oven, separate	4,000 to 5,000
Range	8,000 to 14,000
Range top, separate	4,000 to 8,000
Roaster, large	1,440
Rotisserie (broiler)	1,200 to 1,650
Toaster	500 to 1,200
Toaster oven	1,550
Trash compactor	1,250
Waste disposer	500 to 900

FOOD STORAGE	Watts
Freezer, household	300 to 500
Refrigerator, non-frostless	150 to 300
Refrigerator, frostless	400 to 600

MOTORS	Watts
1/4-hp	300 to 400
1/2-hp	450 to 600
Over 1/2-hp, per hp	950 to 1,000

PERSONAL CARE	Watts
Hair dryer	350 to 1,600
Heating pad	50 to 75
Shaver	8 to 12
Toothbrush	20 to 50

OFFICE EQUIPMENT	Watts
Computer, monitor, printer	200 to 600
Fax machine	200

ENVIRONMENTAL COMFORT	Watts
Air conditioner, central	2,500 to 6,000
Air conditioner, room	800 to 2,500
Blanket	150 to 200
Dehumidifier	250
Fan, attic	800
Fan, bathroom ventilator	500
Fan, ceiling	40 to 300
Fan, whole house	1,000
Fan, portable	50 to 200
Heat lamp (infrared)	250 to 500
Heater, portable	1,000 to 1,500
Heater, wall mounted	1,000 to 4,500
Heater, water bed	800
Heating, electric forced air	10,000
Humidifier	450

ENTERTAINMENT	Watts
Compact disc player	10 to 25
Projector, slide or movie	300 to 500
Stereo	30 to 100
TV (up to 46 in)	150 to 260
VCR	40 to 70

LAUNDRY	Watts
Dryer	4,000 to 6,000
Iron	600 to 1,200
Washer	300 to 800
Water heater	2,000 to 5,000

MISCELLANEOUS	Watts
Clock	2 to 3
Hot tub (electrically heated)	12 kW
Lamps, fluorescent	7 to 60
Lamps, incandescent	10 upward
Sewing machine	60 to 90
Spa (electrically heated)	12 kW
Swimming pool heater	12 kW
Vacuum cleaner	250 to 1,200

TYPES OF ELECTRIC CURRENT

Current is either alternating or direct. On batteries, one terminal is always positive (+) and the other terminal is negative (–). The type of current where each wire is always of the same polarity—either positive or negative—is known as direct current (dc). Current from a battery is always dc. The current coming into your home or farm is alternating current (ac).

Alternating current (ac)   In alternating current each wire changes or alternates continually between positive and negative. The change from positive to negative and back again to positive is known as a "cycle." This takes place 60 times every second, and such current is known as 60-hertz or 60-Hz current. The term "hertz" is used instead of "cycles per second." It is named for the German scientist Heinrich Hertz who discovered the cyclical nature of electrical waves.

Sixty times every second each wire is positive, and 60 times every second it is negative, and 120 times every second there is no voltage at all on the wire. The voltage is never constant but is always gradually changing from zero to a maximum of usually about 170 volts, but averaging 120 volts, and such current is known as 120-volt current (see Fig. 3–3).

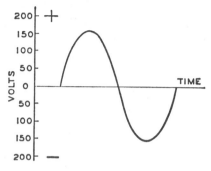

You might wonder why lights do not flicker if there is no current in the wire 120 times every second. The filament in the lamp does not cool off fast enough to create flickering, but very small lamps used on 25-Hz current (where there is no voltage on the wire 50 times every second) do have an annoying flicker.

Fig. 3–3 One cycle of 120-volt alternating current. If it is 60-Hz current, all the changes shown take place in $\frac{1}{60}$ second.

Single-phase and three-phase current   The current described in the previous paragraph is single-phase current. Remember that if 120-volt 60-Hz current flows in a pair of wires, 120 times every second the wires are dead (no voltage at all); 120 times every second the voltage is about 170 volts; at all other times it is somewhere in between, but averaging 120 volts. The voltage is always changing, but the changes are so rapid that for most purposes it can be considered a steady 120-volt current. Single-phase power may be supplied with either two or three wires from the power supplier's transformers (see Fig. 7–1 and related text).

Three-phase power is used in factories and commercial establishments where there are many motors. It is seldom found in homes or farms. To understand how three-phase power is generated, imagine three separate electric generators all on a single shaft, arranged so the voltage reaches its maximum at different times in each of the three generators: first in one, then in the second, then in the third, then again in the first, and so on. A pair of wires would run from each of the three generators. The three generators together then are said to deliver three-phase power (although the power from any one generator is still single-phase). In actual practice, the three generators become a single generator with three coils of wire (windings); the three pairs of wires become three wires.

Three-phase power requires three-wire instead of two-wire high voltage transmission lines and three transformers instead of one. (Three-phase power may be supplied with either three or four wires from the transformers.) Do not be misled into thinking that because there are three wires in the service entrance, the result must be three-phase power. On farms and in homes, the presence of three wires almost always means single-phase three-wire 120/240-volt service. If you are fortunate enough to have three-phase power available, by all means use three-phase motors for greater efficiency. If you do have a 240-volt three-phase supply, consult the power supplier to determine whether you also have 120/240 volt single-phase power for lighting and small appliances without installing your own transformer.

CHAPTER 4
Wire—Selecting and Connecting

WIRES ARE USED as conductors. A conductor is any material that can carry the flow of electric current. The terms *wire* and *conductor* are used interchangeably in this book. *Insulators* are materials that do not conduct electric current. Metal wire is enclosed in insulating material such as plastic to help protect against stray current.

Electricity flows more easily in some materials than others. Copper wire is the best material for ordinary purposes. If iron wire were used, it would have to be about ten times as large in cross-sectional area as copper wire. Other conductors include cable and busbars. Humans and animals can accidentally become conductors resulting in electric shock, so always use caution when working with electricity.

All references in this book are to copper wire, except in the discussion of aluminum wire later in this chapter and in Chapter 19.

WIRE SIZES

Copper wire sizes are indicated by number—the larger the number, the smaller the wire. See Fig. 4–1. The most common size for house wiring is No. 14, which is not quite as big as the lead in a pencil. Nos. 12, 10, 8 and so on are larger than No. 14; Nos. 16, 18, 20 and so on are progressively smaller. No. 14 is the smallest size permitted for ordinary house wiring, and No. 1 is generally the heaviest used in residential and farm wiring. Still heavier sizes are Nos. 1/0, 2/0, 3/0 and 4/0, the 4/0 being almost half an inch in diameter. No. 16 and No. 18 conductors are used mostly in flexible cords and the still finer sizes are used mostly in the manufacture of electrical equipment such as motors. No. 18 is also commonly used in wiring doorbells, chimes, thermostats and similar items operating at less than 30 volts.

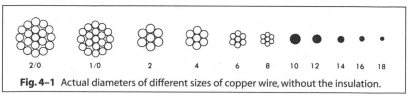

Fig. 4–1 Actual diameters of different sizes of copper wire, without the insulation.

Wire of the correct size must be used for safety and efficiency as discussed below in terms of *ampacity* and *voltage drop.*

Ampacity   Ampacity is the safe carrying capacity of a wire as measured in amperes. When current flows through wire, it creates a certain amount of wasted heat. The greater the amperes flowing, the greater the heat. Doubling the amperes without changing the wire size increases the amount of heat four times. To avoid wasted power, a wire size that limits the waste to a reasonable figure should be used. Of even more concern, if the amperage is allowed to become too great, the wire may become hot enough to damage the insulation or even cause a fire. The National Electrical Code (NEC) is not concerned with wasted power, but it is concerned with safety; therefore it sets the ampacity—the maximum amperage that various sizes and types of wires are allowed to carry.

Table 4–1 AMPACITY OF COPPER WIRES
Columns A and B are based on the NEC Table 310-16; columns C and D are based on NEC Table 310-17; columns E and F are based on NEC Table 310-21.

WIRE SIZE	WIRE TYPE	Not more than 3 conductors in conduit, cable, or buried directly in the earth		Single conductors in free air			
		A	B	C	D	E	F
		TW	RH, RHW, THW	TW	RH, RHW, THW	bare copper	weatherproof
14		15*	15*	15*	15*	30	30
12		20*	20*	20*	20*	35	35
10		30*	30*	30*	30*	50	50
8		40	50	60	70	98	103
6		55	65	80	95	124	130
4		70	85	105	125	155	163
2		95	115	140	170	209	219
1/0		125	150	195	230	282	297
2/0		145	175	225	265	329	344
3/0		165	200	260	310	382	401

* In the NEC, the ampacities of these wires are shown higher than this, but a footnote reference to NEC Section 240-3 limits their overcurrent protection to the figures given here.

Conservative ampacities for copper wire are shown in Table 4–1. For aluminum wire, consult NEC Tables 310-16 and 310-17. Insulation types are described under the "Wire Types" heading in this chapter. Equipment and devices (circuit breakers, switches, panelboards, etc.) have been tested, unless otherwise marked, at the ampacities in Column A of Table 4–1 up to 100 amperes and in Column B over 100 amperes, so even though the NEC tables might indicate the conductor has a higher ampacity, the termination may limit the ampacity to those shown.

Conductors have different ampacities for each variation in ambient temperature, proximity to other conductors, type of insulation, depth of burial, etc. Variations in

WIRES, CIRCUITS, AND GROUNDING

ambient temperature are covered in the NEC by the correction factors at the bottom of NEC Tables 310-16 through 310-19. Adjustments to ampacity due to numbers of conductors in the same raceway or cable are covered in NEC Section 310-15(b)(2). NEC Section 310-15(c) gives a formula for calculating, under engineering supervision, conductor ampacities for conditions not covered in NEC Tables 310-16 through 310-19.

Voltage drop If forcing too many amperes through a wire only caused a certain amount of wasted power, we might consider it a mere nuisance and minor loss. However, it also causes voltage drop. Actual voltage is lost in the wire so that the voltage across two wires is lower at the end than at the starting point. For example, if you connect two voltmeters into a circuit, as in Fig. 4–2, one at the main switch and one across a 1-hp motor at a distance, you will find that the voltage at the motor is lower than at the main switch. The meter across the main switch may read 120 volts. If No. 14 wire is used to the motor, the voltage across the motor terminals will be about 119 volts if the motor is 10 feet away, but only about 112 volts if it is 100 feet away. The difference is lost in the wire and is known as voltage drop. Voltage drop is wasted power, but there is another important consideration: appliances work very inefficiently on voltages lower than the voltage for which they were designed.

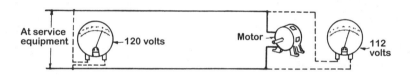

Fig. 4–2 This illustrates voltage drop. The voltage at the motor is lower than the voltage at the starting point.

At 90 percent of rated voltage, a motor produces only 81 percent of normal power and a lamp produces only 70 percent of its normal light.

In the example shown in Fig. 4–2, if 200 feet of No. 14 wire is used, the drop is from 120 to 112 volts, or 8 volts, about 7 percent. If No. 12 wire had been used, the drop would have been reduced about 60 percent to 3.2 volts, only about 2.5 percent of the starting voltage. The larger the wire, the less the voltage drop.

Voltage drop can't be reduced to zero, but it can be kept at a practical level by using wire of sufficient size. A drop of 2 percent is considered acceptable. If the starting point is 120 volts, 2 percent is 2.4 volts, so the actual voltage at the point where the power is consumed is 117.6 volts. If the starting point is 240 volts, the voltage at the point of consumption is 235.2 volts. The apparent saving in initial cost by using undersize wire is soon offset by the cost of power wasted in the wires and by the reduction in efficiency of lamps, motors, and so on.

Selecting wire size   Choose a type and size of wire that has an ampacity rating at least equal to the expected load (in amperes) and that is large enough to limit voltage drop to a practical range. The NEC permits nothing smaller than No. 14 for ordinary wiring. It is better to consider No. 12 the smallest, as this is required in a few places by local ordinance. Larger size wires may be required to minimize voltage drop in long runs, or to allow larger motors to start (wire size for motors is further discussed on page 164). If you need wire heavier than the minimum permitted, it is somewhat complicated to figure the right size but simple to look it up in tables. First determine the amperage to be carried by the wire. Most appliances will have a nameplate giving the total amperes or watts rating.

Tables 4–2 and 4–3 on the next page show one-way distances in feet for 2 percent voltage drop at 120 and 240 volts. These are recommendations in NEC Sections 210-19(a) and 215-2(d). Use the table that corresponds to the voltage of the circuit in question. To operate a load 300 feet away requires 600 feet of wire, but look for the figure 300 in the table. The distances under each wire size are the distances that size wire will carry the different amperages (or wattages) in the left-hand columns, with the customary 2 percent voltage drop. In the 120-volt table, to determine how far No. 8 wire will carry 20 amps, follow the 20-amp line until you come to the No. 8 column: the answer is 90 feet. If a distance is marked with an asterisk (*), it indicates that Type TW wire in conduit or cable, or a cable buried directly in the ground, may not be used because the amperage in the left-hand column is greater than the ampacity of Type TW. Select the proper type of wire from Table 4–1 on page 26, or from Tables 310-16 through 310-19 in the NEC.

Compare the 120-volt and 240-volt tables. Note that at 240 volts, any given size of wire will carry the same amperage twice as far as at 120 volts with the same percentage of voltage drop. It will carry the same number of watts four times as far.

When wires are run outdoors overhead, they must be large enough to carry the amperage involved without excessive voltage drop. They must also be large enough to support their own weight. NEC Section 225-6(a) requires a minimum of No. 10 for spans up to 50 feet and No. 8 for larger distances. For distances over 150 feet, it is wise to use an extra pole. In northern areas where the wires often must support a heavy ice load, consider using a size larger than electrically required. If the wire is installed on a hot summer day, leave considerable slack; otherwise the contraction of the wires in winter may pull the insulators off your buildings.

Weatherproof wire has a covering of neoprene or impregnated cotton over the conductor. The NEC calls it "covered"—the covering is not recognized as insulation. Weatherproof wire must not be used for ordinary wiring, but only for overhead wiring outdoors. Its ampacity is shown in Column F in Table 4–1 on page 26. Size for size, the ampacity of weatherproof wire is higher than other wires, and it may last longer under exposure to the weather—but remember that the smaller the wire, the greater the voltage drop.

Table 4–2 ONE-WAY DISTANCES FOR 2% VOLTAGE DROP BY WIRE SIZE AT 120 VOLTS SINGLE-PHASE

Amperes	Watts at 120 volts	No. 14	No. 12	No. 10	No. 8	No. 6	No. 4	No. 2	No. 1/0	No. 2/0	No. 3/0
5	600	90	140	225	360	570	910				
10	1,200	45	70	115	180	285	455	725			
15	1,800	30	45	70	120	190	300	480	765	960	
20	2,400	20*	35	55	90	145	225	360	575	725	915
25	3,000	18*	28*	45	70	115	180	290	460	580	730
30	3,600	15*	24*	35	60	95	150	240	385	485	610
40	4,800			28*	45	70	115	180	290	360	455
50	6,000			23*	36*	55	90	145	230	290	365

In the tables above and below, the figures represent ONE-WAY distances in feet, not the total wire length for two-way distances.

* In both tables, for distances marked with an asterisk (*) Type TW wires in conduit or cable may not be used because they do not have enough ampacity. For all distances marked with the asterisk, select a type of wire with sufficient ampacity (depending on whether in conduit or cable, or in free air) from Table 4–1 on page 26.

If you wish to permit 4% drop, double the distances shown. Multiply the distances by 2.5 to permit 5% drop.

Table 4–3 ONE-WAY DISTANCES FOR 2% VOLTAGE DROP BY WIRE SIZE AT 240 VOLTS SINGLE-PHASE

Amperes	Watts at 240 volts	No. 14	No. 12	No. 10	No. 8	No. 6	No. 4	No. 2	No. 1/0	No. 2/0	No. 3/0
5	1,200	180	285	455	720	1,145					
10	2,400	90	140	225	360	570	910	1,445			
15	3,600	60	95	150	240	380	610	970	1,530		
20	4,800	45*	70	115	180	285	455	725	1,150	1,450	
25	6,000	35*	55*	90	140	230	365	580	920	1,160	1,460
30	7,200	30*	48*	75	120	190	300	480	770	970	1,220
40	9,600		36*	56*	90	140	230	360	575	725	915
50	12,000			45*	70*	115	185	285	460	580	725
60	14,400				60*	95*	150	240	385	485	610
70	16,800				50*	80*	130	205	330	410	520
80	19,200					70*	115*	180	285	360	460
90	21,000					60*	100*	160	250	320	405
100	24,000					55*	90*	145*	230	290	365
125	30,000						75*	120*	190	240	300
150	36,000							95*	150*	195*	245
200	48,000							70*	115*	145*	185*

WIRE TYPES

Wire of the correct type as well as of the correct size must be used to assure a safe installation. Wires covered with various types of insulation are used for wiring the interiors of buildings. (Wires for outdoor installations are discussed at the bottom of page 28 and on page 175 in the chapter on farm wiring.) Wire names indicate the type of insulation, and these names are generally abbreviated.

The different colors of wires indicate function. The special uses of white and green wires are summarized on page 48 and pages 60–61. Black wires are "hot" (energized)—they carry current to electrical equipment. Red and blue, if used, are also hot; use of such additional colors makes it easier to tell wires apart in a complicated installation. No. 4 and larger wires are usually available only in black, and weatherproof wire in all sizes is always black.

No. 10 and smaller wire is usually solid—the copper conductor is a single solid strand. No. 8 may be solid if it is in the form of cable or is not to be drawn into conduit after installation. But No. 8 that is to be drawn into conduit must be stranded—several smaller wires are grouped together to make one larger wire (see Fig. 4–1) which is more flexible. Wire that is No. 6 or larger (with the exception of weatherproof wire) must always be stranded.

To protect wires from damage they are pulled into pipe called conduit or are used in the form of multi-conductor cables. In addition to the wires described in this book, many other types are available but they are usually not used in residential and farm wiring. You will find them listed in your copy of the NEC.

Fig. 4–3 Type TW is the most common wire in use. It may be used in wet or dry locations. No. 6 or larger must be stranded, and No. 8 must be stranded if pulled into conduit.

Types TW and THW wire   The most commonly used wire is thermoplastic insulated. The conductor is covered by a single layer of plastic compound, the thickness of which depends on the size of the wire, and which strips off easily and cleanly. See Fig. 4–3. Type TW is moisture-resistant and suitable for use in wet locations. Type THW is resistant to both heat and moisture. Neither TW nor THW may be buried directly in the ground.

Type R wire   Formerly rubber-covered, this kind of wire has a synthetic polymer insulation, and it may have a moisture-resistant, flame-retardant outer covering of neoprene or PVC. Fig. 4–4 shows the makeup of the original Type R conductor.

Fig. 4–4 Rubber-covered wire has rubber instead of plastic insulation, and it may have a fabric or other nonmetallic flame-retardant outer covering.

Once the most popular of all kinds of wire, Type R is used comparatively little today. Available are Type RH (75°C) and RHH (90°C) for dry and damp locations, and RHW (75°C) and RHW-2 (90°C) for dry or wet locations. In sizes 8 and larger the NEC assigns a slightly higher ampacity to Type R than to Type TW.

Aluminum wire This is available in two types—aluminum and copper-clad aluminum (aluminum with a thin sheath of copper on the outside). Because aluminum is a relatively poor conductor, a larger size must be used than when using copper. A rule-of-thumb is to use aluminum two numbers heavier than copper: No. 12 aluminum instead of No. 14 copper; No. 4 aluminum instead of No. 6 copper, etc.

When aluminum wire was first used, it was connected to ordinary terminals that were suitable for copper, but it soon became evident they were not suitable for aluminum. The connections heated badly and led to loose connections, excessive heating, and sometimes fires. UL then required redesigned terminals, marked AL-CU, that were considered suitable for either copper or aluminum, but in the 15-amp and 20-amp ratings they were still not suitable for aluminum. The terminals were further redesigned and since 1971 only devices with the marking CO/ALR are acceptable. Note that terminals rated *higher than* 20 amps were not changed, so those marked AL-CU are still acceptable and they *must* be used when aluminum wire is installed.

The AL-CU or CO/ALR marks are stamped into the mounting yokes of switches, receptacles, and similar devices in order to remain visible without removal from the boxes in which they are installed. On larger equipment the marks are located so they remain visible after installation.

If your existing installation was made using aluminum, you would be wise to inspect all your receptacles and switches. If they are not marked CO/ALR, replace them all with devices that do have the mark, or reconnect them with copper pigtails. (Replacement instructions are on pages 196–197.)

Keep in mind that if the aluminum wire is copper-clad, any kind of terminal may be used. Push-in terminals, shown in Fig. 4–16, may be used with copper or copper-clad aluminum, but *not* with ordinary aluminum.

CABLE TYPES

Wires are often assembled into cables such as nonmetallic-sheathed cable or armored cable. When a cable contains two No. 14 wires, it is known as 14-2 (fourteen-two) cable; if it has three No. 12 wires, it is called 12-3, and so on. If a cable has, for example, two No. 14 insulated wires and a bare uninsulated grounding wire, it is known as "14-2 with ground." If a cable has two insulated wires, one is always white and one black; if it contains three, the third is red.

Nonmetallic-sheathed cable   This is a very common type of cable containing two or three Type THHN or THHW wires. Many people call it "Romex," which is the trade name of one particular manufacturer. It is easy to install, neat and clean in appearance, and less expensive than other kinds of cable. Brief descriptions of the two kinds follow here. See Chapter 11: *Cable—Nonmetallic-sheathed and Armored* for a complete discussion of where and how to use nonmetallic-sheathed cable.

One kind of nonmetallic-sheathed cable is called Type NM by the NEC. As shown in Fig. 4–5A, the individually insulated wires are enclosed in an overall plastic jacket. (In older construction, occasionally still seen, the outer jacket was a braided fabric. In any case, the outer jacket must be moisture-resistant and flame-retardant.) Some manufacturers put paper wraps on the individual wires or over the assembly. Empty spaces between wires are sometimes filled with jute or similar cord. This type may be used only in normally dry locations, but never in barns on farms.

Fig. 4–5A Nonmetallic-sheathed cable consists of two or more individual wires assembled into a cable. The Type NM two-wire with ground shown here may be used only in dry locations.

Fig. 4–5B Nonmetallic-sheathed cable, Type NMC, may be used in dry or damp locations.

The other kind is called Type NMC by the NEC and is especially designed for damp or corrosive locations such as barns. It may also be used in ordinary dry locations. The individual insulated wires in Type NMC Cable are embedded in a solid sheath of plastic material (see Fig. 4–5B). Sometimes there is a glass overwrap on each insulated wire. There is no fibrous material such as paper wraps or jute filler that can be affected by moisture as in ordinary Type NM.

Armored cable   NEC Type AC (Fig. 4–6) is called "BX" by many people although that is the trademark of one particular manufacturer. The individual wires may be type TW, THW, or THHN. There may be an overwrap of tough paper between the

Fig. 4–6 Armored cable consists of two or more individual wires assembled into a cable and protected by flexible steel armor. It may be used only in dry locations.

wires and the spiral steel armor. Where installed in thermal insulation, the conductors are required to be rated 90°C but with 60°C ampacity (making it unnecessary to go through the math for derating for high ambient temperature as required at the bottom of NEC Table 310-16). It is for use only in permanently dry locations. Details for using armored cable are provided in Chapter 11.

Underground cable   Two special types of cable, USE and UF, are used for underground wiring. They are described in the chapter on farm wiring on page 175.

FLEXIBLE CORDS

Flexible cords are used to connect lamps, appliances, and other loads to outlets. Each wire consists of many strands of fine wire for flexibility. Over the wire is a wrapping of cotton to prevent the insulation from sticking to the copper. There are many kinds of flexible cords with varying kinds and thicknesses of insulation depending on the purpose of the cord. The more common kinds are described here.

Type SPT-2 is the most common cord used for radios, floor lamps, and similar loads. As shown in Fig. 4–7, it consists of copper wires embedded directly in plastic insulation. It is tough, durable, and available in various colors. The same kind of cord with a rubber-like insulation is known as Type SP-2. These cords are commonly available only in No. 18 and No. 16.

Fig. 4–7 In Type SPT-2, the wires are embedded in plastic. The cord is durable and attractive.

Fig. 4–8 Types S and SJ cords are designed for very heavy use.

Type S, in Fig. 4–8, is a really durable cord that stands up to hard use. Each wire has rubber or plastic insulation. The two wires are bundled into a round assembly with jute or paper twine filling the empty spaces. Over all is a layer of tough, high-grade, rubber-like plastic. Type SJ is similar but with a thinner outer layer. Type SV or SVT are very similar but more flexible cords used on vacuum cleaners.

If the outer jacket is made of oil-resistant thermoset plastic, the cord becomes oil-resistant and the designations become SO and SJO instead of S and SJ.

"Heater cord" is used for irons, toasters, and other heating appliances. Heater cords formerly had asbestos under a cotton braid as part of the insulation system. Asbestos is no longer used as wire insulation. Fig. 4–9 shows Type HPN heater cord; the wires are embedded in neoprene.

Fig. 4–9 Type HPN cord is used on irons, toasters, and other heating appliances.

TOOLS USED FOR WIRING JOBS

Tools needed for carrying out various electrical jobs are mentioned in the text where the procedures are described. Look in the index under "Tools" or under individual tool names to find page numbers. The tools listed below are commonly used in wiring procedures.

- Cable cutter for cutting armored cable
- Cable ripper for removing the outer jacket from nonmetallic-sheathed cable
- Conduit bender for bending conduit
- Hacksaw for cutting armored cable, EMT, and openings in existing walls
- Putty knife sharpened to lift floorboards to get at ceiling spaces from above
- Keyhole saw for cutting floorboards
- Fish tape for pulling wires into place
- Parallel-jaw pliers for use with connectors
- Pliers with crimping die for use with connectors
- Side-cutting pliers for removing insulation from wire
- Wire stripper or knife for removing insulation from wire
- Test light for receptacles and fuses
- Continuity tester for testing that hot wires in switches are not accidentally grounded and for detecting broken wires in cords

TERMINALS FOR CONNECTING WIRES TO DEVICES

Various kinds of terminals provide the means for connecting wires to devices. Before wires can be connected to a device or spliced to another piece of wire, the insulation must be removed. The procedure begins with preparing the wire.

Removing insulation from wire   Using a tool called a wire stripper is the most convenient way to remove insulation from wire. Another method is to use a pair of side-cutting pliers. Insert the wire close behind the hinge of the plier blades and mash the insulation, softening it, from the point where it is to be removed to the end. Then place the jaws of the pliers at the point where the removal is to begin, squeezing just hard enough to cut into the insulation without touching the conductor; the mashed insulation can then be pulled off easily.

If you don't have a wire stripper or side-cutting pliers, you can use a knife. Do not cut the insulation off sharply, as shown at A of Fig. 4–10, because it is too easy to accidentally nick the conductor, leading to later breaks. Hold your knife to produce an angle as shown in B of Fig. 4–10.

A   **B**

Wrong method   **Right method**

Fig. 4–10 Wrong and right methods of removing insulation from wires.

Always make sure the stripped end of wire is absolutely clean. Rubber-covered wires have a tinned conductor to make it easy to strip off the insulation. Plastic-insulated wires are not tinned because that kind of insulation strips off cleanly.

Two types of terminals There are two kinds of terminals. One kind consists of a terminal screw in a metal part with upturned lugs to keep the wire from slipping out from under the screw, as shown in Fig. 4–11. It may be used with No. 10 and smaller wires, but it is very difficult to make a good connection if the wire is No. 10 and stranded. The other kind is used mostly for wires larger than No. 10; the wires are inserted into the terminal of a solderless connector and the screw is then tightened. See Fig. 4–12.

Insulation

Fig. 4–11 Terminal on a typical receptacle or switch.

Fig. 4–12 Solderless connectors of this type are used with heavy sizes of wire. *(The Halex Company)*

The correct method for terminals for No. 10 and smaller wires is shown in Fig. 4–13. Wrap the wire at least two-thirds (preferably three-quarters) of the way around the screw in a clockwise direction so that tightening the screw tends to close the loop rather than open it. Tighten the screw until it makes contact with the wire, then tighten it about another half-turn to squeeze the wire a bit. As an alternative, tighten the screws to the pound-inches marked on the product, or to 12 pound-inches as recommended in Fig. 4–13. Never make the errors shown in Fig. 4–14. The insulation should end not more than ¼ inch from the screw at the most. Note that these two illustrations were made specifically for aluminum wire, but the principles are also correct for copper wire.

Do not connect two wires under a single wraparound terminal screw even though it might appear logical in some situations. The NEC prohibits it in Section 110-14(a). Take those two wires and another short length of the same wire, and connect all three together using a wire connector described later in this chapter. See Fig. 4–15. Then connect the remaining end of the short wire under the terminal screw. Most connectors of the kind shown in Fig. 4–12 are for one wire unless the connector is marked to indicate the number and size of wires that can be accommodated. The marking may be on the carton if the connector itself is too small for the marking.

What length should the bare wire be for connecting it under the terminal screw? Professionals leave about 3 inches of bare wire, enough to leave a "tail" beyond the tightened screw. Twist this tail a few times and it will break off near the screw. Another way is to leave just enough bare wire to go around the screw, form a loop with a pair of long-nose pliers, slip it around the screw, close it with the pliers so that the loop is entirely under the screw, and then tighten the screw.

Push-in clamps Many 15-amp and 20-amp receptacles and switches have no terminal screws at all. Instead, there are internal terminal clamps that grasp a straight piece of wire pushed into them, forming an effective connection. Merely strip the end of the wire for half an inch or so (the proper length is usually shown on the

Fig. 4–13 Be sure wire is wrapped around terminal screw in clockwise fashion as in Step 1 so that tightening the screw tends to close the loop. Then complete Steps 2 and 3. *(UL)*

STEP 1: STRIP AND WRAP WIRE

STEP 2: TIGHTEN SCREW TO FULL CONTACT

STEP 3: COMPLETE CONNECTION

Fig. 4–14 Avoid these common errors when connecting wire to terminal screws. *(UL)*

WIRES, CIRCUITS, AND GROUNDING

Fig. 4–15 Use this method, commonly called a "pigtail" splice, when you would otherwise have to connect two wires under a terminal screw.

Fig. 4–16 Strip the wire, push it into opening on switch or receptacle, and the connection is made.

◄ **Fig. 4–17** In conduit wiring, a continuous wire may be connected to a terminal as shown here.

device itself), and push the wire into the hole on the device. See Fig. 4–16. If an error is made, release the clamp by pushing a small screwdriver blade into another opening on the device. These push-in connections are acceptable only for copper or copper-clad aluminum wire. They must not be used with all-aluminum wire. For receptacles these connections are limited to No. 14 wire.

Connecting conduit   In conduit wiring, sometimes a wire is pulled through one box and on to another, possibly through still another, and so on. If the wire merely runs through the box, pull it through without splice but leave a loop in case it is necessary to re-pull it in the future. If you intend to make a connection to the wire as it passes through the box, let a loop several inches long project out of the box. Strip away an inch or so of insulation, form a loop, and connect it under one terminal screw as shown in Fig. 4–17.

CONNECTORS FOR SPLICING WIRES

Connectors are used in splicing (joining) two or more pieces of wire together. It is important that whichever type of wire connector you use, you are certain it is Listed for the number and size of wires that are to be joined. The spliced wires must be electrically as good as an unbroken length of wire. The insulation of the splice must be as good as that on the original wires. Such a splice is accomplished by using properly installed insulated solderless connectors. Many people call these "wire nuts," which is the trade name of a particular manufacturer. When two or more ends of wire must be connected to each other, lay the wires together with their cut ends pointing in the same direction; insert the wires into the connector and turn it onto the wires, which twists them together as shown in Fig. 4–18.

Fig. 4–18 Ordinary splicing is simple when using solderless connectors.

Connectors for joining two or more wire ends   One type of connector has a threaded metal insert molded into the insulating shell. Screw the connector onto the wires to be joined. The other kind has a removable metal insert. Slip the insert over the wires to be joined, tighten the screw of the insert, and then screw the insulating shell over the metal insert.

The spring-loaded connector shown in Fig. 4–19 is also popular. Inside its insulating shell there is a cone-shaped metal spring. Screw the connector over the wires to be joined. The insulating cover provides a good grip. When being screwed on, the coil spring temporarily unwraps. When released, it forms a very tight grip on the wire.

Fig. 4–19 This kind of connector contains a tapered coil spring inside the insulating cover.

In using these connectors, if one wire is much smaller than the others, let it project a bit beyond the heavier wires. If you have removed the right length of insulation from each wire, the insulating shell will cover all bare wires and no taping is necessary. Note that all types of these connectors are available in various sizes, depending on the number and size of the wires to be joined.

For wire sizes No. 10 and smaller, the insulated "clamshell" connector shown in Fig. 4–20 may be used. The wire insulation is used to position the wire in this type of connector, so do not strip the wire insulation before installing in the connector. A squeeze with parallel-jaw pliers installs the self-insulating connector.

Another wire connector, for No. 10 and smaller wire, is the shell shown in Fig. 4–21. Unless the manufacturer's instructions direct otherwise, first twist the wires together, then slip the shell over them and crimp, using a tool of the type shown. Then cut off the wire ends that could puncture the insulation, and either tape or use formed plastic caps to insulate.

Fig. 4–20 "Clamshell" wire connector for taps or pigtails. For copper wire only.

Fig. 4–21 Crimp-type wire connector, and pliers with special crimping die in handle. For copper wire only.

Fig. 4–22 For heavier wires, use metal connectors. The assembled connector and wires must be taped.

For wires that are too large to be joined by the connectors described above, use heavy duty copper connectors of the style shown in Fig. 4–22.

Connectors for splicing to a continuous wire   Sometimes a wire must be spliced to another continuous wire. In the heavier sizes, the simplest way is to use one of the split-bolt connectors shown in Fig. 4–23. Tape after making the connection. Some connectors are available with an insulating cover that can be snapped on after making the connection. For smaller sizes, it is usually simpler to cut the continuous wire to form two ends; the wire to be spliced in the connection makes the third wire. Then use a solderless connector as shown in Fig. 4–15.

Fig. 4–23 Use a split-bolt connector when splicing a heavy wire to another continuous heavy wire. The assembled connector and wires must be taped. *(The Halex Company)*

Insulating splices   Splicing devices such as solderless connectors are self-insulating. For other styles, insulating covers or boots are available that can be added after the splice is made. Some must be taped (see Figs. 4–22 and 4–23). Use electrician's plastic tape—it has a very high insulating value despite its thinness. Starting well back on the wire insulation, wrap the tape on spirally from one end to the other, partially overlapping each turn, and then return with the spiral in the opposite direction. If the splice will be subject to mechanical strain or abuse, apply additional layers as a cushion.

GETTING STARTED—THE BASICS
▶ **WIRES, CIRCUITS, AND GROUNDING**
INSTALLING SERVICE EQUIPMENT AND WIRING
SPECIAL WIRING SITUATIONS AND PROJECTS

CHAPTER 5
Circuit Protection and Planning

OVERCURRENT DEVICES such as fuses and circuit breakers limit the amperage in any wire to the maximum that is permitted by the National Electrical Code (NEC). If more than the permissible maximum amperage is allowed to flow, the temperature of the wire goes up and the insulation may be damaged, leading to shortened life and accidental grounds that can become dangerous. If the overload is great enough, there is danger of fire. The maximum amperage established by the NEC as safe for any particular kind and size of wire is called the ampacity. Table 4–1 lists the ampacity of copper wires.

OVERCURRENT DEVICES

Any overcurrent device you use must have a rating in amperes not greater than the ampacity of the wire that it protects. For example, if No. 12 wire has a calculated ampacity of 20 amps, the circuit breaker or fuse that you use to protect the wire must have a rating not greater than 20 amps.

When two different sizes of wire are joined together (for example, when No. 8 is used for mechanical strength in an overhead run to a building where it is joined to No. 14 for the inside wiring), the overcurrent protection must be the right size for the smaller of the two wires. A fuse or circuit breaker of the correct size for the larger wire may be used at the starting point, provided another one of the proper size for the smaller wire is used where the wire is reduced in size. An example is when a wire such as No. 8, with an ampacity of 40, runs from one building to another where it feeds several 15-amp circuits (see Fig. 5–1).

Fig. 5–1 If fuses are used where the wire size is reduced, select a fuse size that protects the smaller wire.

What do you do when a fuse blows or a circuit breaker trips? Most people will say to install a new fuse or reset the breaker. That's correct—but first find out what caused the blown fuse or tripped breaker. Fuses and circuit breakers are the safety valves of electrical installations. Using substitutes or fuses too large for the size of wire can be dangerous because it could lead to a fire. Chapter 19 explains how to troubleshoot a blown fuse or tripped circuit breaker.

Fuses A fuse is nothing more or less than a short piece of metal of a kind and size that will melt when more than a predetermined number of amperes flows through it. This metal link is enclosed in a convenient housing to prevent hot metal from spattering if the fuse blows and to permit easy replacement. A fuse rated at 15 amps is tested to carry 15 amps. When more than 15 amps flow through it, the wire inside the fuse melts (the fuse "blows"), which is the same as opening a switch or cutting the wire. The greater the overload, the quicker the fuse will blow.

Plug fuses The common plug fuse, shown in Fig. 5–2, is made in ratings up to 30 amps. It is known as the Edison-base type because the base is the same as on ordinary lamps. These fuses are not permitted in new installations. They may be used only as replacements, and then only when there is no evidence of tampering or overfusing (using a fuse too large for the size of wire involved). For new installations, see the entry for Type S nontamperable fuses. Plug fuses are rated at 125 volts, but may be used on a system having a grounded neutral and no conductor over 150 volts to ground, so they could be used for a 240-volt load served from a 120/240 volt three-wire system.

Fig. 5–2 Edison-base plug fuses are made only in ratings up to 30 amps. They are not permitted in new installations.

Time-delay fuses When a motor is started, fuses often blow because a motor that draws only 6 amps while running may draw as much as 30 amps for a few seconds while starting. An ordinary fuse carries 80 percent of its rated amperage indefinitely, but blows very quickly if twice that amperage flows through it. However, wire that can safely carry 15 amps continuously—but might be damaged or even cause a fire if 30 amps flowed continuously—will not be damaged or cause a fire if 30 amps flow for a few seconds. The "time delay" type of fuse takes advantage of the ability of wire to carry a higher load momentarily. Commonly called by the trade name "Fusetron" although there are other brands, it looks like an ordinary fuse but is made differently inside. It blows just as quickly as an ordinary fuse on a small continuous overload or on a short circuit, but it will carry a big overload safely for a fraction of a minute. This type of fuse is convenient where motors are used because it prevents needless blowing of fuses and eliminates many service calls. Edison-base time-delay fuses are not permitted in new installations. They are permitted as replacements only, and then only if there is no evidence of overfusing or tampering. For new installations, see the entry for Type S nontamperable fuses.

WIRES, CIRCUITS, AND GROUNDING

Type S nontamperable fuses The NEC requires the use of Type S fuses in all new installations that use fuses and for replacements if there is evidence of tampering or overfusing. Since all ratings of ordinary Edison-base plug fuses are interchangeable, nothing prevents someone from using, for example, a 25-amp or 30-amp fuse to protect a No. 14 wire, which must be protected at not over 15 amps. To prevent this unsafe practice, the nontamperable fuse was developed (see Fig. 5–3). The NEC calls this a "Type S" fuse, but it is commonly called by the trade name "Fustat." The fuse itself will not fit an ordinary fuseholder, so an adapter

Fig. 5–3 A typical Type S nontamperable fuse (center, and cross section at right) and its adapter (left). Once an adapter has been screwed into a fuseholder, it cannot be removed. This prevents the use of fuses larger than originally intended.

(shown at left in the illustration) must first be installed in the ordinary fuseholder; once installed, it cannot be removed. There are three sizes of adapters. As a safety measure, the 15-amp will accept only 15-amp or smaller fuses; the 20-amp will accept only 16-amp through 20-amp fuses; the 30-amp will accept 21-amp through 30-amp fuses. The Type S fuse is presently made only in the time-delay type. They are rated at 125 volts, but may be used on a system having a grounded neutral and no conductor over 150 volts to ground, so they could be used for a 240-volt load served from a 120/240 volt three-wire system.

A word of caution when using Type S fuses—when screwing the fuse into its holder, turn it some more after it appears to be tight. Under the shoulder of the fuse there is a spring—the fuse must be screwed in tightly enough to flatten the spring or the fuse will not "bottom" and you will have an open circuit just as if the fuse were blown.

Cartridge fuses This type is made in all amperage ratings. Those rated at 60 amps or less are of the ferrule type shown in Fig. 5–4. Those rated at 70 amps or more have knife-blade terminals as shown in Fig. 5–5. The cartridge fuses of the type illustrated may be used in any circuit of up to 250 volts between conductors. Cartridge fuses are used for loads of 30 amps and higher such as mains, dryers, ranges, resistance heaters, and water heaters.

Fig. 5–4 Cartridge fuses rated 60 amps or less are of the ferrule type shown.

Fig. 5–5 Cartridge fuses rated more than 60 amps have knife-blade terminals as shown.

Circuit breakers   In new residential construction, circuit breakers are generally installed, with fuses appearing mostly in existing installations. A circuit breaker looks like a toggle switch. Fig. 5–6 shows a single unit. Inside each breaker is a fairly simple mechanism which in case of overload trips the breaker and disconnects the load. If a breaker trips because of overload, in most brands you must force the handle beyond the OFF position, then return it to ON to reset it. On some brands, the handle merely goes to the OFF position and is reset it by returning it to the ON position.

A circuit breaker has a definite time delay. It will carry 80 percent of its rated load indefinitely, a small overload for a considerable time, and will trip quickly on a large overload. Nevertheless, it will carry temporary overloads long enough to permit motors to start.

Fig. 5–6   A single circuit breaker (left) and method of resetting the breaker if it trips.

CALCULATING BRANCH CIRCUIT NEEDS

If all the lights and appliances in a home or on an entire farm were protected by a single fuse or circuit breaker, the entire establishment would be in darkness when that breaker tripped or the fuse blew. Also, all the wires would have to be very large to match the ampere rating of that breaker or fuse, which would make a clumsy and expensive installation. Therefore the different outlets in an installation are separated into smaller groups known as branch circuits. Inexpensive No. 14 or No. 12 wire is used for most of the wiring, protected by 15-amp or 20-amp breakers or fuses. When one of these breakers trips or a fuse blows, only those outlets on that circuit are dead; those on other circuits are still live.

Continuous loads   See the discussion under "Water heaters" on page 152 in Chapter 14: *Appliances.*

Outlets per circuit   In residential work, the NEC does not limit the number of lighting outlets placed on one circuit. But if you put too many, you will probably have trouble with breakers tripping or fuses blowing. In most cases, it is best to connect fewer than a dozen outlets on one circuit even if more are permitted.

Note: An outlet is any point where electric power is actually used. Each fixture, even if it has five lamps, is considered one outlet. Each receptacle outlet, even the duplex type, is one outlet. Switches are not outlets since they use no electric power, but merely control its use. But in estimating the cost of an installation, switches are included in counting the outlets to arrive at a total cost on the "per outlet" basis, though this is not in accordance with the NEC definition.

Kinds of circuits   The circuits used in homes can be divided into the following three general types: *lighting circuits*—primarily for lighting and serving permanently installed lighting fixtures, as well as receptacle outlets into which you plug lamps, radios, televisions, clocks, and similar 120-volt loads (but not kitchen appliances); *small appliance circuits*—receptacles in kitchen, dining room, etc. for such items as coffeemaker, toaster, electric fry pan; *individual appliance circuits*—each serving a single appliance such as a range, water heater, clothes washer or dryer; circuits may be either 120-volt or 240-volt.

In addition, there may be a need to add a circuit for a permanently connected motor that is not part of an appliance, such as might be used with certain shop tools. (Motor types and installation are discussed in Chapter 16.)

Lighting circuits The NEC requires enough lighting circuits to provide 3 VA of power for every square foot of floor space in the house. Since a circuit wired with No. 14 wire and protected by 15-amp overcurrent protection provides 1,800 VA ($15 \times 120 = 1,800$), each circuit is enough for 600 square feet ($1,800 \div 3 = 600$). This is the NEC *minimum*. The NEC concerns itself with safety only. To allow for convenience and usefulness over time, it is wise to provide one circuit for each 500 square feet of space.

What is included in the square footage of a house when calculating the number of lighting circuits needed? Open porches and garages, even if attached to the house, are not included in the total. But unfinished or unused spaces that are adaptable for future use must be included according to the NEC. Run a special circuit to such an area, terminate it in a single outlet if you wish, and later branch off from that outlet to the additional outlets that you want when you finish off the space. If the space is large, run two circuits.

To arrive at the NEC definition of the total number of square feet for determining the number of circuits, multiply the length of the house by its width. If its outside dimensions are 30×45 feet, then one floor represents 1,350 square feet ($30 \times 45 = 1,350$). If the basement is finished or can be finished into usable space, add its area of, for example, 24×30 or 720 square feet to the first-

Table 5–1 NUMBER OF LIGHTING CIRCUITS		
USABLE AREA	CODE MINIMUM	RECOMMENDED
1,200 sq ft	2	3
1,600 sq ft	3	4
2,000 sq ft	4	4
2,400 sq ft	4	5
2,800 sq ft	5	6
3,200 sq ft	6	7

floor area of 1,350 square feet for a total of 2,070 square feet. If an upper floor has unfinished space that can later be finished into a bedroom, add its area.

Then divide the total area by 600 to arrive at the minimum number of lighting circuits the NEC requires, or divide by 500 for a more adequate installation. In Table 5–1, minimum and recommended numbers of lighting circuits are calculated for some typical square footages.

Special small-appliance circuits   The circuits already discussed are for lighting, including floor and table lamps, and items such as radios, TVs, and vacuum cleaners. These circuits do not permit proper operation of larger kitchen and similar appliances that consume much more power. NEC Section 210-52(b) requires two special small-appliance circuits to serve only small-appliance outlets, including refrigeration equipment, in the part of the house occupied by kitchen, pantry, breakfast room, and dining room. Both circuits must extend to the kitchen; the other rooms may be served by either or both of them. The circuits must be wired with No. 12 wire protected by 20-amp circuit breakers or fuses. No lighting outlets may be connected to these circuits (with two exceptions: a combination receptacle/support for an electric clock in kitchen or dining area, and a receptacle serving the ignition system, oven light, and/or timer for a gas-fired cooking appliance). Either 15-amp or 20-amp 125-volt rated receptacles may be installed on these 20-amp circuits.

Each such circuit has a capacity of 2,400 VA ($20 \times 120 = 2,400$), which is not too much considering that toasters, irons, and similar appliances often require 1,000 watts, and appliances such as roasters and toaster ovens may consume over 1,500 watts. These two circuits can be merged into one three-wire circuit, which is discussed on pages 108–110.

At least one separate 20-amp circuit must be run to laundry appliances, and another one to all the bathroom receptacles. A separate 15-amp branch circuit is permitted to serve refrigeration equipment.

Individual circuits   It is customary to provide a separate circuit for each of the following:

- Self-contained range
- Separate oven, or counter-mounted cooking unit
- Water heater
- Clothes washer
- Clothes dryer
- Waste disposer
- Dishwasher
- Motor on oil-burning furnace
- Motor on blower in furnace
- Water pump
- Permanently connected appliances rated at more than 1,000 watts (example: a bathroom heater)
- Permanently connected motors rated more than $1/8$ hp

Appliance and motor circuits may be either 120-volt or 240-volt depending on the particular item installed. These circuits are discussed in more detail in Chapter 14: *Appliances*. See pages 195–196 for instructions on how to label every circuit at the panelboard to assist in troubleshooting.

Motor circuits Separate circuits are recommended for individual motors over ⅛ hp that are not part of an appliance. Such motors might be either directly connected (example: sump pump) or belt-driven (example: bench grinder). Determine the motor's amperage from Table 5–2, adapted from NEC Table 430–148. See also the discussion on page 162.

Total number of circuits You must decide for yourself how many circuits you will need. Provide circuit breakers or fuses for specific appliances even if you do not intend to install such appliances until later. Even then, provide for some spare circuits. For a new wiring installation, follow the

Table 5–2 DETERMINING CORRECT AMPERAGE FOR MOTORS		
MOTOR	120 VOLTS	240 VOLTS
¼ hp	6 amps	3 amps
⅓ hp	7 amps	3 ½ amps
½ hp	10 amps	5 amps
¾ hp	14 amps	7 amps
1 hp	16 amps	8 amps
1 ½ hp	20 amps	10 amps
2 hp	24 amps	12 amps
3 hp	34 amps	17 amps
5 hp	56 amps	28 amps

suggestions in Table 5–3 for calculating the total number of 120-volt and 240-volt circuits needed based on the square footage of usable area.

Table 5–3 SUGGESTED NUMBER OF 120-VOLT AND 240-VOLT CIRCUITS							
	Square feet of area in house ▶	1,200	1,600	2,000	2,400	2,800	3,200
CALCULATE 120-VOLT CIRCUITS	General-purpose circuits (one per 500 sq. ft.)	3	4	4	5	6	7
	Spare general-purpose circuits	1	1	2	2	2	2
	Small appliance	2	2	2	2	2	2
	Special laundry	1	1	1	1	1	1
	Bathroom receptacles	1	1	1	1	1	1
	Individual appliance circuits:						
	Oil or gas burner	1	1	1	1	1	1
	Blower on furnace	1	1	1	1	1	1
	Bathroom heater, room air conditioner, workshop motor, etc.	1 each	1 each	1 each	1 each	1 each	1 each
	Spares	1	1	2	2	2	2
	Fill in total 120-volt circuits ▶						
CALCULATE 240-VOLT CIRCUITS	*Individual appliance circuits:*						
	Range	1	1	1	1	1	1
	Water heater	1	1	1	1	1	1
	Clothes dryer	1	1	1	1	1	1
	Water pump, other fixed in place	1 each	1 each	1 each	1 each	1 each	1 each
	Spares	1	1	1	1	1	1
	Fill in total 240-volt circuits ▶						

CHAPTER 6

Circuit Diagrams

BEFORE YOU CAN WIRE a building, you must learn how switches, receptacles, sockets and other devices are properly connected to each other with wire to make a complete electrical system called a circuit—the path along which current flows.

GROUNDED WIRES

In residential and farm wiring one of the current-carrying wires is grounded, which means it is connected to an underground metal water pipe system and to a driven ground rod. Many people refer to the grounded wire as the neutral wire. Sometimes it is a neutral and sometimes it is not. (Chapter 7 contains a full discussion of grounding and the Index lists specific grounding topics.) Ungrounded wires are known as "live" or "hot" or "energized" wires, which means they have a voltage—in house wiring, usually 120 volts—above zero (ground potential—the earth is assumed to be at zero potential).

It is essential to remember these five points about grounding:

- The grounded wire is white in color. (Two recognized alternatives to white for the grounded conductor are: natural gray; or three white stripes on any color insulation except green.)

- The grounded wire must run direct to every 120-volt device or outlet to be operated (never to anything operating only at 240 volts).

- The grounded wire is never fused or protected by a circuit breaker.

- The grounded wire is never switched or interrupted in any other way.

- White wire must never be used except as a grounded wire. Other wires are usually black but may be some other color, but not white or green.

The white wire must run to every 120-volt device or outlet other than a switch; the other wire to the device is usually black but may be some other color, but not white or green. Examine a socket or similar device carefully and you will find that one of its two terminals is a natural brass color and the other is silver-colored (usually nickel-plated, and called "white" by the NEC). The white wire must always run to the white terminal. In the case of sockets, the white terminal is always connected to the screw shell and never to the center contact of the socket. Switches never have white terminals.

WIRING DIAGRAMS

Wiring diagrams are used in planning electrical installations. Diagrams are useful for planning safe and efficient installations, and for clarifying details such as the wiring of three-way and four-way switches.

Fig. 6-1 This symbol indicates a lamp in all diagrams.

Fig. 6-2 This method is used in all diagrams to indicate whether or not crossing wires are connected to each other.

Crossing wires not connected   **Crossing wires connected together**

Diagram symbols   For any circuit, two wires must run from the starting point to each outlet serving electrical equipment (lamp, motor, etc.). Imagine that the current flows out over the black wire to the equipment, through the equipment, and back to the starting point over the white wire. In this book, the starting point is called the SOURCE, and a lamp symbol (Fig. 6-1) is used to indicate the equipment to be operated. Fig. 6-2 illustrates the method used to indicate whether or not crossing wires are connected to each other. In all diagrams wires are indicated like this:

white wire   black wire   black wire between a switch and
                                            the outlet which it controls

The fact that one line is heavier than the other does not mean that one wire is larger than the other; both are the same size.

Adding switches   The simplest possible diagram is that of Fig. 6-3, a lamp that is always on with no way of turning it off. Such a circuit is of little value, so you should add a switch. Fig. 6-4 shows how you simply make the black wire detour to a switch. An unenclosed switch is shown so you can see how it operates: opening (turning off) the switch is the same as cutting a wire. Of course, unenclosed switches are unsafe and not permitted. Fig. 6-5 shows the same circuit but with an enclosed flush toggle switch which serves the same purpose, but is much neater, more convenient, and safer to use.

Fig. 6–3 The simplest circuit—
a lamp always on.

Fig. 6–4 The circuit of Fig. 6-3 plus an
open switch.

Fig. 6–5 A toggle switch (closed)
substituted for the open switch in
Fig. 6–4.

Wiring in parallel If several lamps are used, do not make the mistake of wiring
them *in series* as in Fig. 6–6 because that is impractical for most purposes. If one
lamp burns out or is removed from its socket, it is the same as opening a switch in
the circuit—all the lamps go out. Instead, wire them *in parallel* as in Fig. 6–7. The
white wire goes to each lamp, from the first to the second and to the third and so on;
the black wire also runs to each lamp in similar fashion. Each lamp will light even if
one or more of the others is removed. The switch shown cuts off the current to all
the lamps at the same time. The wiring is simple: white wire from SOURCE to each
socket or outlet; black wire from SOURCE to the switch, and from the switch another
black wire to each of the sockets. This is the way a fixture with five sockets is wired.

Fig. 6–6 Style of wiring known as
"series" wiring. It is impractical
except for very special purposes.

Fig. 6–7 Style of wiring known as
"parallel" wiring. It is used for all
ordinary purposes.

WIRES, CIRCUITS AND GROUNDING

If each lamp is to be controlled by a separate switch, use the Fig. 6–8 diagram. The white wire always runs from source to each lamp; the black from source to each switch, and from each switch a black wire runs to the lamp that it controls. If you trace the current from the source over the black wire through each switch, one at a time, along the black wire to the lamp, and back from the lamp to the source, you will find that each lamp can be independently controlled.

Fig. 6–8 This diagram shows the same parallel wiring as Fig. 6-7, but with each lamp controlled by a separate switch. Note that the white wire runs directly from source to each lamp.

Wiring receptacle outlets The wiring of plug-in receptacle outlets is simple. If there is only one receptacle, run the white wire from the source to the side of the receptacle that has white (silver-colored) terminal screws, and run the black wire to the other side, as in Fig. 6–9. If there are several receptacles, as in Fig. 6–10, run the white wire to one side of the first receptacle, from there to the second receptacle, and so on; do the same with the black wire. Fig. 6–10 is the same parallel wiring diagram as Fig. 6–7 except that receptacles have now been substituted for lamps and the switch has been omitted.

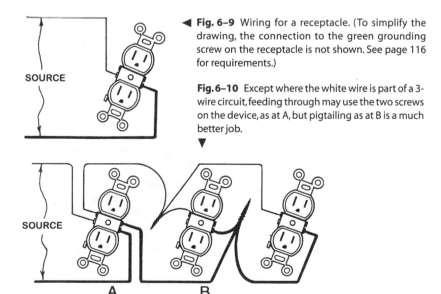

◀ **Fig. 6–9** Wiring for a receptacle. (To simplify the drawing, the connection to the green grounding screw on the receptacle is not shown. See page 116 for requirements.)

Fig. 6–10 Except where the white wire is part of a 3-wire circuit, feeding through may use the two screws on the device, as at A, but pigtailing as at B is a much better job.

▼

A   B

Where the two wires shown in Fig. 6–10 are part of a three-wire circuit (see pages 108–110 for three-wire circuit installation), it is not permitted to feed the white conductor through by using the two screws on the side of the receptacle, as shown at A, because removing this receptacle could place 120-volt loads in series on 240 volts, with possibly disastrous results. Instead, a splice is required with a pigtail going to the receptacle as shown at B. Many inspectors require pigtailing at all feed-through receptacles even though the devices are tested and listed to be connected as shown at A. (See Fig. 4–15 for another illustration of a pigtail splice.)

Wiring outlets controlled by pull chains   There are few locations where a pull chain will be used, since lighting in attics and under-floor spaces used for storage or for equipment requiring servicing must be controlled by a wall switch at the entry to the space. To wire for a pull chain, substitute a pull-chain socket (shown at F in Fig. 9–4) for the receptacle in Fig. 6–9.

Combining several diagrams   When several groups of outlets are to be wired, the two separate wires from each group can be run back to a common starting point where the wires enter the house, as shown in Fig. 6–11 where the diagrams of Figs. 6–5, 6–7, and 6–8 have been combined. This method requires a considerable amount of material because the wires must be much longer than necessary. It is more efficient just to wire the first group, then run wires from the first group to the second, and from the second to the third, as in Fig. 6–12 which shows the same three groups but using less material.

The white wire may be extended at any point to the next outlet. The black wire likewise may be extended at any point, provided that it can be traced all the way back to the SOURCE without interruption by a switch. In other words, in these diagrams a black wire can be extended from any black wire indicated by the solid heavier lines, but not from the black wire between a switch and an outlet indicated by a heavy broken line. Thus, in Fig. 6–12, A and B are the starting points for the second group, and C and D are the starting points for the third group of lights.

Three-way switches   The switch used for controlling a light from one point is known as a "single pole" switch. "Three way" switches allow you to control a light from two points (not three, despite the name). With three-way switches you can turn a hall light on or off from upstairs and downstairs, or a garage light from house and garage, or a yard light from house and barn. Such switches have three different terminals for wires. Their internal construction is similar to Fig. 6–13. In one position of the handle, terminal A is connected inside the switch to terminal C; in the other position, terminal A is connected to terminal B. Usually, the common terminal A is identified by being a darker color than the other terminals which are natural brass.

Study the diagram of Fig. 6–14 on page 54 in which the handles of both switches are down. The current can be traced from SOURCE over the black wire through Switch 1, through terminal A, out through terminal B and up to terminal C of Switch 2, but there it stops. The light is off.

Fig. 6–11 The outlets of Figs. 6-5, 6-7, and 6-8 have been combined into a single group of outlets. In other words, here they form one circuit. This still uses a great deal of material. To reduce the amount required, the circuit can be arranged as the next diagram shows.

Fig. 6–12 The diagram of Fig. 6-11 rearranged to use less material.

Fig. 6–13 This shows what happens inside a 3-way switch when the handle is thrown from one position to another. The current enters by the common terminal A, and leaves by either C or B depending on the position of the handle.

Fig. 6–14 A light controlled by two 3-way switches. With both handles down, the light is off.

Fig. 6–15 With the handles of both 3-way switches up, the light is still off.

Next see Fig. 6–15 where the handles of both switches are up. The current can be traced as before from SOURCE over the black wire through Switch 1, through terminal A, but this time out through terminal C, and from there up to terminal B of Switch 2. There it stops and the light is off.

Now examine the diagrams of Figs. 6–16 and 6–17. In both diagrams the handle of one switch is up, the other down. In each case the current can be traced from SOURCE over the black wire, through both switches, through the lamp, and back to SOURCE. The light in both cases is on. In the case of either Fig. 6–16 or Fig. 6–17, throwing either switch to the opposite position changes the diagram back to either Fig. 6–14 or Fig. 6–15, and the light is off. In other words, the light can be turned on and off from either switch.

The wiring of three-way switches is simple, as these diagrams show. Run the white wire from SOURCE as usual to the light to be controlled. Run the black wire from SOURCE to the common or marked terminal of the first three-way switch. Run a black wire from the common or marked terminal of the second three-way switch to the light. That leaves two unused terminals on each switch; run two black wires from the terminals of the first switch to the two terminals of the second switch. It makes no difference whether you run a wire from B of the first switch to C of the

Fig. 6-16 Three-way switch: with one handle down and the other up, the light is now on. Trace the current along the arrows.

Fig. 6-17 Three-way switch: both handles are now the opposite of Fig. 6-16, but the light is still on. Either switch controls it.

second as shown, or from B of the first to B of the second. The wires that start at one switch and end at another are called runners, travelers, or jockey legs.

Four-way switches   To control a light from more than two points, use two three-way switches, one nearest the source and the other nearest the light, and four-way switches at all remaining points in between. Four-way switches can be identified by the fact that they have four terminal screws and do not have on-off markings on the handles. Figs. 6-18 through 6-21 show two types of four-way switches and the wiring diagrams for circuits using each type. Depending on the manufacturer, the internal connections of four-way switches may be as shown in Fig. 6-18 or as in Fig. 6-20. In each of these figures, the four terminals are labeled A, B, C, and D.

The four-way switch in Fig. 6-18 is shown in a circuit in Fig. 6-19. The two jockey legs from one three-way switch are connected to terminals A and B of the four-way switch; from the other three-way, the jockey legs are connected to terminals C and D of the four-way.

The four-way switch in Fig. 6-20 is shown in a circuit in Fig. 6-21. The jockey legs from one three-way switch are connected to terminals A and D of the four-way switch, and the jockey legs from the other three-way are connected to terminals B and C of the four-way switch.

If you are not sure which type of four-way switch you have, try one diagram; if it does not work, try the other. You can do no harm by wrong connections except that the circuit will not work if you have the wrong connection.

◀ **Fig. 6–18** Internal connections of one style of 4-way switch with the handle in the up and the down positions. The terminals are labeled A, B, C, and D.

Fig. 6–19 Four-way switches are usually connected as shown here. With the 4-way switch handles in the positions shown in the upper diagram, the light is off; in the lower diagram, the light is on. Two 4-way switches are shown in the circuit, but any number may be installed depending on need. Refer to Fig. 6–18 to identify the terminals.

▼

WIRES, CIRCUITS AND GROUNDING

Fig. 6–20 Internal connections of the other style of 4-way switch with the handle in the up and the down positions. Terminals are labeled A, B, C, and D.

Fig. 6–21 Some brands of 4-way switches are connected as shown. With the 4-way switch handles in the positions shown in the upper diagram, the light is off; in the lower diagram, the light is on. Refer to Fig. 6–20 to identify terminals.

▼

Pilot lights   When a light can't be seen from the switch that controls it, the light is often left on unnecessarily. The problem often occurs with basement and attic lights in homes and haymow lights in barns. Install a pilot light (a small, low-wattage lamp) near the switch so that both lights are turned on and off at the same time.

Fig. 6-22 The basic circuit of a pilot light.

Fig. 6-23 The wiring scheme of a combination pilot light and switch. Another brand might be wired somewhat differently.

The pilot light serves as a reminder that the unseen light is on. The wiring is shown in Fig. 6–22. First look at the diagram while disregarding the wires shown in dotted lines as well as the pilot light itself. It is then the same as Fig. 6–5. Add the wires shown in dotted lines and it becomes the same as the diagram of Fig. 6–7 except there are two lamps instead of five. The white wire runs to both lamps; the black wire from the SOURCE runs to the switch as usual; the black wire from the switch runs to both lamps, and the diagram is finished.

Fig. 6-23 shows a combination switch and pilot light in one device. The diagram is correct only for some brands; exact diagrams usually come with each device. There are also switches having a small pilot light in the handle. Some have a light that is on while the switch is in the OFF position to aid in finding the switch in the dark.

Other diagrams and combinations   You will have little difficulty making a diagram for any desired combination of outlets and switches, whether single-pole or three-way, if you remember the principles covered in this chapter. To make any diagram, first locate each outlet where desired and run the white wire (light line like this ——————) from SOURCE to every outlet. Then run a black wire (heavy line like this ▬▬▬▬▬) from SOURCE to each outlet that is not controlled by a switch. Next run black wire (heavy broken line like this ▬ ▬ ▬ ▬) from each outlet controlled by a switch to the switch that controls it. If three-way switches are involved, run additional black wires between the switches that control the outlet. Finally, from each switch run black wire (heavy line like this ▬▬▬▬) back to SOURCE. Having done that, you may have to rearrange some of the wires to reduce the amount of wire used, but this will not affect the proper operation of the hook-up.

CHAPTER 7

Grounding for Safety

ELECTRICITY ALWAYS FOLLOWS the shortest possible path to the earth, that is, to the ground. In discussions concerning electrical wiring you will regularly meet the terms *ground, grounded,* and *grounding.* They all refer to deliberately connecting parts of a wiring installation to the earth. Actually, the connection is made to something that serves as the earth—a metal rod driven deeply into the ground and metal piping that is part of an underground water system.

The purpose of grounding is safety. A wiring installation that is properly grounded allows excess electrical current—such as from lightning strikes—to travel into the earth without causing serious injury to people or damage to the wiring system. Grounding facilitates the proper functioning of fuses and circuit breakers, limiting the risk of shock from defective equipment.

An installation that is not properly grounded can present an extreme danger of shock, fire, and damage to appliances and motors. For example, suppose a 2,400-volt line accidentally falls across your 120/240-volt service during a storm. If the system is not grounded, you can be subject to deadly 2,400-volt shocks, and wiring and appliances will be ruined. If the system is properly grounded, the highest voltage of a shock will be much more than 240 volts, but very much less than 2,400 volts. Lightning striking on or even near a high voltage line can cause great damage to your wiring and your appliances, and it can cause fire and injuries. Proper grounding throughout the system greatly reduces the danger.

Because grounding is so important it is discussed repeatedly throughout this book. Be certain that you understand it thoroughly. The National Electrical Code (NEC) rules for grounding are extensive and sometimes seem ambiguous. But for installations in homes and farm buildings—the only installations covered in this book—the rules are relatively simple. This chapter discusses the basic principles of grounding. The practical details, including installing ground rods and ground clamps and selecting the correct ground wire size, are covered in the next chapter on installing the service entrance.

HOW DANGEROUS ARE SHOCKS?

Most people think it is a high voltage that causes fatal shocks. This is not necessarily so. The amount of current flowing through the body determines the effect of a shock. A milliampere is one-thousandth of an ampere (0.001 amp). A current of 1 milliampere through the body is just barely perceptible. Currents from 1 to 8 milliamperes cause mild to strong surprise. Currents from 8 to 15 milliamperes are unpleasant, but usually the victim is able to let go and get free of the object that is causing the shock. Currents over 15 milliamperes are likely to lead to "muscular freeze," which prevents the victim from letting go and often leads to death. Currents over 75 milliamperes are almost always fatal; much depends on the individual involved.

The higher the voltage, the higher the number of milliamperes that would flow through the body under any given set of circumstances. A shock from a relatively high voltage while the victim is standing on a completely dry surface will result in fewer milliamperes than a shock from a much lower voltage to someone standing in water. Deaths have been caused by shock from circuits considerably below 120 volts, while someone standing on a dry surface could survive shock from a circuit of 600 volts and more. How to help the victim of a shock is discussed on page 7.

GROUNDING BASICS

This section defines grounding terminology and describes the functions of the wires and equipment in your installation. Italics are used to indicate words for which definitions are provided and to help differentiate among similar-looking terms.

Ground   For the wiring installations covered in this book, the term "ground" means underground metal water piping connected to a driven metal rod forming a continuous conductive path that allows excess current to travel into the earth. The rod driven into the earth is called a ground rod. The water pipe is part of the building's underground water system. The NEC calls the rod and water pipe "grounding electrodes." Properly joined together, they form a "grounding electrode system." This arrangement is illustrated at the bottom of Fig. 8–6. When it is said that something is "grounded," it means that it is connected to ground.

Grounding falls into two categories. System grounding is for the current-carrying wires in an installation. Equipment grounding is for parts that do not carry current in the installation—they are listed under "Grounding (green) wire" on the next page.

Bonding (or bonded)   Bonding means connecting normally non-current-carrying metal parts (conduit, boxes, etc.) to each other and finally to *ground*, usually using bare uninsulated wire, resulting in a conductive path that will safely carry accidental voltage into the earth.

Ground wire   This is the wire that runs from the service equipment to the grounding electrode (which is grounded because it is buried in the earth). The NEC calls it the "grounding electrode conductor." It is usually bare, but can be insulated and of any color but green. The ground wire is bonded within the service

WIRES, CIRCUITS, AND GROUNDING

equipment enclosure to the neutral conductor, to the service raceway, to the service enclosure, and to the equipment grounding bus if any.

Grounded (white) wire   In a circuit, this is the wire (usually white) that normally carries current and is connected to the *ground* at the service equipment. The NEC calls it the "identified conductor." The grounded wire must never be fused or protected by a circuit breaker or interrupted by a switch.

Grounding (green) wire   This is a wire that does not carry current at all during normal operation. The NEC calls it the "equipment grounding conductor." It is *bonded* to components of the installation that normally do not carry current but do carry current in case of damage to or defect in the wiring system or the appliances connected to it. These components include the equipment grounding bus in the service equipment cabinet (which holds the main fused service switch or the circuit breaker panelboard, whichever is installed), frames of motors, frames of appliances such as electric range and clothes washer, the outlet boxes in which switches or receptacles are installed, and the metal conduit or the armor of armored cable.

The grounding wire runs with the current-carrying wires. It must be green, green with one or more yellow stripes, or bare. In this book it will be called simply a green wire. It must never be used for any purpose except as the grounding wire in a circuit. In the case of wiring with metal conduit, or cable with a metal armor, a grounding wire as such is not installed separately because the conduit or the armor of the cable serve as the grounding conductor where it is bonded to the cabinet.

GROUNDING OVERVIEW

Proper grounding involves every part of the installation, including the incoming service wires that run from the power supplier's transformer to the building to be served. The service equipment cabinet—the heart of your electrical installation—is where the main grounding connections are made.

Wires from transformer to building   Fig. 7–1 shows the power supplier's three wires—labeled A, B, and N—that run from the transformer to the building. The grounded wire, labeled N, is grounded both at the transformer and at the building's service equipment. This is a neutral wire (meaning it is non-current-carrying when the load is balanced on the hot wires). Wires A and B are "hot" wires. (Hot wires carry voltage above zero, or ground. One is usually black and the other red, or both black, but never white or green.) The voltage between A and N, or between B and N, is 120 volts; between A and B it is 240 volts.

Fig. 7–1 The power supplier's three wires provide two different voltages. Use the lower voltage (120 V) for low-wattage loads, and the higher voltage (240 V) for high-wattage loads such as range and water heater. Note the symbol for a connection to ground.

Between the transformer and the service equipment, the grounded wire N is a neutral wire. In wiring a building, any wire connected to the point where the neutral wire ends in the service equipment is simply a grounded wire. Many people call it a neutral but it is not a neutral; there cannot be a neutral in a two-wire circuit. (Three-wire circuits, which have a grounded neutral wire, are discussed on pages 108–110.)

Grounding of 120-volt and 240-volt loads The grounded wire must run without interruption to all equipment operating at 120 volts, but not to anything operating only at 240 volts. Only hot wires run to 240-volt loads. A separate grounding wire runs to 240-volt loads (unless the conduit or the armor of armored cable serves as the grounding conductor). *Note:* Anything that is connected to a circuit and consumes power constitutes a "load" on the circuit. The load might be a motor, a toaster, a lamp—anything consuming power. Switches do not consume power and therefore are not loads. A receptacle is not a load, but anything plugged into the receptacle is a load.

Grounding at the service equipment At the service equipment all the grounding wires are connected to an equipment grounding busbar, which is bonded to the enclosure and connected to the neutral busbar (for all the white wires), which is in turn connected (grounded) by means of the ground wire to the water pipe and ground rod. The power supplier also grounds the neutral of the incoming service wires at the transformer serving the building. If the neutral wire is properly grounded both at the transformer and at the building, it follows that if you touch an exposed grounded wire at a terminal or splice, no harm follows—no shock— any more than if you touch a water pipe or a faucet, because the grounded wire and the piping are connected to each other. Any time you touch a pipe, you are in effect also touching the grounded wire.

Short circuits and ground faults If two hot wires touch each other at a point where both are bare, or if a hot wire touches a bare point in a grounded circuit wire, a short circuit occurs at the point of connection. This rarely happens in the actual wires of a properly installed system, but often happens in cords to lamps or appliances, especially if the cords are badly worn or abused.

A ground fault results when a bare point in a hot wire, such as where the wire is connected to a receptacle or switch, touches a grounded component such as conduit, the armor of armored cable, or a grounding wire. For both short circuit and ground fault the effect is the same: a fuse will blow or a circuit breaker will trip.

HOW GROUNDING PROMOTES SAFETY

The diagrams in this section illustrate correct installations as well as faulty conditions in which shock can occur. The motors represented could be either free-standing or part of an appliance. The coiled portion of the line on the right side of each diagram represents the motor's winding. Your risk of shock and danger from a faulty installation will depend on the surface on which you are standing, your general physical condition, and the condition of your skin at the contact point. If

WIRES, CIRCUITS, AND GROUNDING

you are on an absolutely dry surface you will note little shock. If you are on a damp surface (as in a basement) you will experience a severe shock. If you are standing in water you will undergo extreme shock or death. As a safety precaution, always stand on dry boards when you must work in a damp or wet location.

Grounding and fuse/breaker placement for 120-volt circuit   Fig. 7–2 shows a 120-volt motor with the grounded wire connected to the grounded neutral of the service equipment and a fuse in the hot wire. (Actually the fuse and the ground connection would be at the service equipment cabinet a considerable distance from the motor, not near the motor as shown, although there might be an additional fuse near the motor.) If the fuse blows, the motor stops. What happens if, in inspecting the motor, you accidentally touch one of the wires at the terminals of the motor? Nothing happens because the circuit is hot only up to the blown fuse. Between the fuse and the motor the wire is now dead just as if the wire had been cut at the fuse location. The other wire to the motor is grounded, so it is harmless. You are protected. But if the fuse is *not* blown and you touch the hot wire, you will receive a 120-volt shock through your body to the earth, and through the earth back to the neutral wire at the service equipment.

◀ **Fig. 7–2** A 120-V motor with grounded wire correctly connected to the grounded neutral and a fuse correctly placed in the hot wire. (Grounding wire not yet installed.)

Fig. 7–3 A 120-V motor with a fuse wrongly placed in the grounded wire. It is a dangerous installation. ▶

Fuse/breaker placement error   The same circuit is shown in Fig. 7–3 except the fuse is wrongly placed in the grounded wire instead of the hot wire. The motor will operate properly. If the fuse blows, the motor stops. But the circuit is still energized through the motor and up to the blown fuse. If you touch one of the wires at the motor, you complete the circuit through your body, through the earth, to the neutral wire in your service equipment; you are directly connected across 120 volts and as a minimum you will receive a shock, and at worst will be killed.

Accidental internal ground—unprotected Now see Fig. 7–4, which again shows the same 120-volt motor as in Fig. 7–2. But suppose the motor is defective so that at the point marked G the winding inside the motor accidentally comes into electrical contact with the frame of the motor. As a result, the winding "grounds" to the frame. That does not prevent the motor from operating. But suppose you choose to inspect the motor, touching just its frame. What happens? Depending on whether the internal ground between winding and frame is at a point nearest the grounded wire or nearest the hot wire, you will receive a shock up to 120 volts as you complete the circuit through your body back to the ground*ed* wire. It is a potentially dangerous situation since shocks of much less than 120 volts can be fatal.

In fact, it is not uncommon for breakdowns in the internal insulation of a motor to result in an accidental electrical connection between the winding and the frame of the motor. The entire frame of the motor becomes hot. The same situation arises if the motor is fed by a cord that becomes defective where it enters the junction box of the motor so that one of the bare wires in the cord touches the frame. If there is no cord and the motor is fed by the circuit wires, a sloppy splice between the circuit wires and the wires in the junction box on the motor can lead to the same result: the frame of the motor becomes hot.

◀ **Fig. 7–4** The same motor as in Fig. 7–2, but the motor is defective. G represents an accidental grounding of the winding to the frame. The grounding wire has not been installed. This is a dangerous installation.

Fig. 7–5 The same defective motor as Fig. 7–4, but now a grounding wire has been installed from the frame of the motor to ground. Even though the winding is accidentally grounded to the frame , as represented by G, there is no shock hazard. ▶

Accidental internal ground—protected Now see Fig. 7–5 which shows the same motor as in Fig. 7–4 with the same accidental ground between winding (or cord) and frame, but protected by a grounding wire that is connected to the frame of the motor and runs back to the ground connection at the service switch. When the internal ground occurs, current will flow over the ground*ing* wire. It will sometimes but not always blow the fuse. Even if the fuse does not blow, the ground*ing* wire will

WIRES, CIRCUITS, AND GROUNDING

protect you because it reduces the voltage between the frame of the motor to substantially zero as compared with the ground you are standing on. You will not receive a shock provided that a really good job of bonding was done at the service. If the bonding is poor, you will still receive a shock.

The ground*ing* wire from the frame of the motor (or from any other normally non-current-carrying component) may be green or in many cases bare uninsulated wire. Green wire may not be used for any purpose other than the ground*ing* wire. Other chapters discuss when a separate ground*ing* wire must be installed. If metal conduit or cable with armor is used, the conduit or armor serves as the ground*ing* wire.

Grounding and fuse/breaker placement for 240-volt circuit   Now refer to Fig. 7–6, which shows a 240-volt motor installed with each hot wire protected as required with a fuse or circuit breaker. Remember that in such 240-volt installations, the white ground*ed* wire does not run to the motor, but is nevertheless grounded at the service equipment. If you touch both hot wires, you will be completing the circuit from one hot wire to the other, and you will receive a 240-volt shock. But if you touch only one of the wires, you will be completing the circuit through your body, through the earth, back to the grounded neutral in your service equipment, and you will receive a shock of only 120 volts: the same as touching the ground*ed* wire and one of the hot wires of Fig. 7–1. The difference between shocks of 120 volts and 240 volts may be the difference between life and death.

◀ **Fig. 7–6** A 240-V motor with each hot wire correctly protected by fuse/breaker. The grounded wire does not run to 240-V loads. Because the grounding wire has not been installed, shock hazard is avoided *only if the motor remains in perfect condition.*

Fig. 7–7 This is the same installation as Fig. 7–6, but the motor is now defective. The winding is accidentally grounded to the frame, as represented by G. This is a dangerous installation. ▶

Accidental internal ground—protected   Assume that the motor in Fig. 7–6 becomes defective, that the winding of the motor becomes accidentally grounded to the frame, as shown in Fig. 7–7. This as the same as Fig. 7–4 except that the motor is operating at 240 volts instead of 120 volts. Touching the frame will produce a 120-volt shock, but if the frame has been properly grounded as in Fig. 7–8, one of the fuses will probably blow. Even if a fuse does not blow, the severity of the shock you receive from touching the frame will be limited because the frame is grounded, again assuming that a really good ground was installed at the service.

GROUNDING CONSIDERATIONS FOR WIRING SYSTEMS

In any of the situations of Figs. 7–2 through 7–8, if there is an accidental contact between any two wires of the circuit, the contact constitutes a short circuit, and a fuse will blow or the circuit breaker will trip regardless of whether the short is between one of the hot wires and the grounded wire, or between the two hot wires. For a short circuit to occur, there must be bare places on two different wires touching each other, which does not happen very often in a carefully installed job.

◄ **Fig. 7–8** The same defective motor as Fig. 7–7, but a grounding wire has been installed from the frame of the motor to ground. Even though the winding is accidentally grounded to the frame, as represented by G, the shock hazard is minimized.

Fig. 7–9 Wires in metallic conduit or metallic armor are used to install the motor. The conduit is grounded both at the service equipment cabinet and to the motor. It is a safe installation. ▶

Advantages of metal conduit or armored cable   Consider a wiring system in which all the wires are installed in a metal raceway: metal pipe called conduit, or cable with a metal armor (commonly referred to as "BX"). The raceway or armor is grounded at the service equipment. It is also connected to the motor itself (assuming that the motor is not connected by a flexible cord and plug). No separate grounding wire is then required. If there is now an accidental ground from winding to frame (or to the raceway or armor), it has the same effect as a short between one of the hot wires and the grounded wire, because the grounded wire and the raceway or armor are connected to each other and to the ground at the service equipment. A fuse will immediately blow whether the motor operates at 120 or 240 volts. A considerable advantage has been gained—accidental grounds that otherwise might remain undiscovered are automatically disclosed. See Fig. 7–9.

Where nonmetallic-sheathed cable is recommended   In spite of the advantages of a metal raceway system, in some locations, especially farm buildings housing livestock, conditions of high humidity and corrosion make the use of a cable without metal armor more desirable (see page 180). Then nonmetallic-sheathed cable Type NMC is used, which in addition to the usual insulated wires also has a bare uninsulated ground*ing* wire running throughout the wiring system.

Continuous grounding   As explained on the first page of Chapter 9, an outlet box or switch box is used to protect every electrical connection in an installation. When a grounded-neutral wiring system is used (which is 100 percent of the time for installations of the kind discussed in this book), and you use metal conduit or cable with a metal armor, you must ground not only the neutral wire but also the conduit or cable armor. The neutral is grounded only at the service equipment (at the point where it is connected to the neutral of the service equipment), but the conduit or armor must be securely connected to every box or cabinet. Lighting fixtures installed on metal outlet boxes are automatically grounded through the conduit or armor. (Nonmetallic boxes are also used; see Figs. 17–1 and 17–2.) Motors or appliances directly connected to conduit or armor are automatically grounded. If using nonmetallic-sheathed cable, the extra wire—the bare uninsulated ground*ing* wire—must be carried from outlet to outlet, providing a continuous ground. The connecctions for this wire are explained on pages 115–116. Regardless of the wiring method used, a continuous ground all the way back to the service is essential.

GROUNDING-TYPE RECEPTACLES INCREASE SAFETY

Many old receptacles still in use have only two parallel openings for the plug, as in Fig. 7–10. Plugging something into this receptacle duplicates the condition of Fig. 7–2. If you handle a defective appliance that is plugged into an old-style receptacle, you could receive a shock (see Fig. 7–4). This danger led to the development of the "grounding receptacle," which has three openings (see Fig. 7–11). Note that the

Fig. 7–10 Polarized receptacle and plug, with the wider slot connecting to the grounded conductor. A polarized plug cannot be inserted into a non-polarized receptacle, so replace an old non-polarized receptacle with the polarized type.

Fig. 7–11 Grounding receptacle and plug. Plugs with either 2 or 3 blades will fit.

grounding receptacle in Fig. 7–11 has the usual two parallel slots for two blades of a plug, plus a third round or U-shaped opening for a third prong on the corresponding plug. In use, the third prong of the plug is connected to a third or grounding (green) wire in the cord, running to and connected to the frame of the motor or other appliance.

Making the ground connection   On the receptacle, the round or U-shaped opening leads to a green terminal screw that in turn is connected to the metal mounting yoke of the receptacle. The details of how to connect the green terminal to the equipment grounding conductor, or to the grounded metal box, are discussed in Chapter 11 on pages 115–116. In this way, the frame of the motor or appliance is effectively grounded, leading to extra safety as shown and discussed in connection with Fig. 7–5.

Where grounding receptacles are required   If an appliance is connected by cord and plug, NEC Section 250-114(3) requires a three-wire cord and three-prong plug on every refrigerator, freezer, air conditioner, clothes washer, clothes dryer, dishwasher, sump pump, aquarium equipment, and every hand-held, motor-driven tool such as a drill, saw, sander, hedge trimmer and similar items. The three-wire cord with three-prong plug is not required on ordinary household appliances such as toasters, irons, radios, TVs, razors, lamps and similar items.

Only one kind of receptacle needs to be installed, since grounding receptacles are designed for use with both two-prong and three-prong plugs. The NEC requires that in all new construction only grounding receptacles are to be installed. (See page 198 for replacing two-wire receptacles in existing installations.)

GROUND-FAULT CIRCUIT INTERRUPTER (GFCI)

The GFCI is a supplementary protection that senses leakage currents too small to operate ordinary branch circuit fuses or circuit breakers. The use of a grounding wire in a three-wire cord with a three-prong plug and a grounding receptacle reduces the danger of a shock in some circumstances, such as when using a portable tool, but it does not eliminate the danger completely. Cords can be defective or wrongly connected. Millions of tools with two-wire cords are still in use. Some people foolishly cut off the grounding prong on a three-prong plug because they have only two-wire receptacles (see Figs. 7–10 and 7–11).

Under normal conditions the current in the hot wire and the current in the grounded wire are absolutely identical. But if the wiring or a tool or appliance is defective and allows some current to leak to ground, then a ground-fault circuit interrupter will sense the difference in current in the two wires. If the fault current exceeds the trip level of the GFCI, which is between 4 and 6 milliamperes, the GFCI will disconnect the circuit in as little as one-fortieth of a second.

The fault current, which is much too low to trip a normal breaker or blow a fuse, could possibly flow through a person in contact with the faulty equipment and a grounded surface. The use of a GFCI is a highly recommended safety precaution, especially when using electrical equipment outdoors where standing on the ground (especially if it is wet) greatly increases the likelihood and especially the severity of a shock. The GFCI you install must be rated in amperes and volts to match the rating of the outlet or circuit it is to protect.

The GFCI should be considered additional insurance against dangerous shocks. It is not to be considered as a substitute for grounding. The GFCI will not prevent a person who is part of a ground-fault circuit from receiving a shock, but it will open the circuit so quickly that the shock will be below levels that inhibit breathing or heart action or the ability to let go of the circuit.

Where GFCIs are required   NEC Section 210-8 requires GFCI protection for 15-amp and 20-amp receptacles in these locations in dwellings: at kitchen counters; within 6 feet of a wet bar sink; outdoors at balconies, porches, patios, decks, and roofs; in detached accessory buildings; in garages (except where not accessible such as a garage door opener or for appliances in dedicated space such as a freezer); in wired detached garages; in unfinished basements; in crawl spaces; and in all boathouses. GFCI protection is also required for bathroom receptacles in all occupancies, and for permanently installed receptacles used temporarily during construction. Some swimming pool, spa, and hot tub installations (not covered in this book) also require GFCI protection.

Three types of GFCI   Ground-fault circuit interrupters are available as separately enclosed types, or in combination with either a breaker or a receptacle.

- The *separately enclosed type* is available for 120-volt two-wire and 120/240-volt three-wire circuits up to 60 amps. It is most often used in swimming pool wiring, installed at any convenient point in the circuit.

- The *breaker type* combines a 15-amp, 20-amp, and up to 60-amp circuit breaker and a GFCI in the same plastic case. It is installed in place of an ordinary breaker in your panelboard, and is available in 120-volt two-wire or 120/240-volt three-wire types (which will protect a 120/240-volt three-wire circuit or a 240-volt two-wire circuit). It provides protection against ground faults and overloads for all the outlets on the circuit. You can at any time replace an ordinary breaker in your panelboard with one of these combination breakers. Each GFCI circuit breaker has a white pigtail that you must connect to the grounded (neutral) busbar of your panelboard. You must connect the white (grounded) wire for the circuit to a terminal provided for it on the breaker.

- The *receptacle type* combines a receptacle and a GFCI in the same housing. It provides only ground-fault protection to the equipment plugged into that receptacle or, if it is the "feed through" type, to equipment plugged in to other ordinary receptacles installed "downstream" on the same circuit. This type is a convenient choice when replacing existing receptacles where GFCI protection is desired or required.

GFCI testing and potential problems Regardless of the type or brand of GFCI you install, it is essential that you carefully follow the installation and periodic testing instructions that come with it. Every GFCI has a test button for easy verification of its functional operation.

The GFCI is designed to trip if the cords or tools plugged into the protected receptacle outlet are in poor repair and provide a path for current to leak to ground. Even where wiring, tools, and appliances are in perfect condition and there is no ground fault, be on the lookout for these installation problems that will cause tripping of a GFCI:

- a two-wire GFCI receptacle (other than an end-of-run type) is connected in a three-wire circuit
- the white circuit conductor is grounded on the load side of the GFCI
- or the protected portion of the circuit is excessively long (250 feet maximum is the rule of thumb—longer circuits may develop a capacitive leakage to ground)

Remember that the GFCI will not prevent shock, but it will render shocks relatively harmless. Also, it will not protect a person against contact with both conductors of the circuit at the same time unless there is also a current path to ground. The GFCI may be pictured as a tiny computer that constantly monitors the current out to a load and back again. The GFCI acts to quickly disconnect the circuit only when the current out to the load and the current returning differ by 0.005 amp (5 milliamperes) or more.

LIGHTNING ARRESTERS

If you live in an area subject to lightning activity, consider installing a lightning arrester at your service location. Lightning damage to building wiring may not be evident at the time of the strike but may show up later. See page 184 for illustration and further discussion.

SURGE SUPPRESSORS FOR COMPUTERS AND OTHER ELECTRONIC EQUIPMENT

With the increasing home use of personal computers and other sensitive electronic equipment on general-use branch circuits, there is a need for suppression of voltage surges. These surges are typically of very short duration. Among the possible causes are the switching off or on of fluorescent lights or motors such as air conditioners, or the switching of major loads by the utility, or distant electrical storms.

Such surges can cause a personal computer to lose data, or cause other troubles for modems, data terminals, word processors, fax machines, electronic cash registers, computerized sewing machines, and other electronic equipment. Even with a surge suppressor in place, it is a good idea to disconnect (unplug) sensitive electronic equipment when there is an electric storm nearby.

Several types of surge suppressors are available including:

- a plug-in unit resembling a cord adapter
- a group of receptacles on a strip with a supply cord
- permanently installed receptacles that include surge suppression in their design
- a unit shaped like a circuit breaker that provides only surge suppression—it is plugged into a two-pole circuit breaker space
- and circuit breakers which, in addition to the usual overcurrent protection features, incorporate a surge suppressor to protect the entire branch circuit

Some of these may include a replaceable module, or a light or buzzer to indicate when the suppressor has failed, or both. The less expensive models degrade after doing their duty a few times and may not offer any protection. There is no way to test them. Most designs shunt the excess energy into the grounding circuit, which may damage other equipment on the same circuit. The best designs incorporate capacitors to store the excess energy instead of shunting it to the ground. Manufacturers can request additional testing by UL, in addition to safety testing, to establish Endurance Grades A, B, and C; Performance Classes 1, 2, and 3; and Modes 1 and 2. The best surge suppressors carry the Class 1, Grade A, Mode 1 classification. You can spend less than $10 to more than $200, and in general you get what you pay for. Be sure the unit you purchase is UL Listed.

CHAPTER 8

The Service Entrance

ELECTRIC POWER FLOWS from the supplier's transformer through a set of three wires to your service entrance where the wires enter the building. This chapter discusses the design and installation of a service entrance for a house. Installations for farm buildings are discussed in Chapter 17 which begins on page 165.

PLANNING YOUR SERVICE ENTRANCE

These are the components of the service entrance. Fig. 8–6 shows them in use.

Outside the building—

- Service head, also called the entrance cap or weather head
- Insulators anchored to the building below the service head—this is where the power supplier's wires end
- Service entrance cable or wires in conduit leading from the insulators to the service head and down to the meter socket
- Meter socket
- Service entrance cable or wires extend from the meter socket into the building

Inside the building—

- Service entrance cabinet that contains the disconnecting means and overcurrent protection—either circuit breakers or fuses in a panelboard
- Ground connection—a ground wire from the incoming service to the busbar in the cabinet and back outside the building to the grounding electrode system

120-volt or 120/240-volt current   The number of wires installed by your power supplier in your service entrance determines the type of current you will have.

Two-wire service entrance   When only two wires are installed, only 120-volt current is available. One of the wires is white (sometimes bare) beginning at the insulators, and is grounded. It is not a neutral wire; there cannot be a neutral in a two-wire circuit. The hot wire is black.

Three-wire service entrance   The standard installation today is the three-wire entrance. When three wires are installed, both 120 volts and 240 volts are available as shown in Fig. 7-1. The voltage between N and A, or between N and B, is 120 volts; between A and B it is 240 volts. The wire N is white (sometimes bare) beginning at the insulators. It is grounded, and up to the service equipment inside the building it is a *neutral* wire. The two hot wires can be any color except white or green, but are usually black and red, or both black.

Service equipment rating   Service circuit breakers are rated at 30, 40, 50, 60, 70, 90, 100, 125, 150, 175, and 200 amps. Service fusible switches are rated at 30, 60, 100, and 200 amps. Equipment larger than 200 amps is available for larger installations. Some power suppliers will furnish a self-contained meter that will handle a 400-amp service, but beyond that the meter must be served by current transformers in a separate enclosure ahead of the service disconnect.

The service equipment of a residence is required to have a rating of 100 amps or higher. In wiring buildings other than houses, a 30-amp service may be installed if there are not more than two 2-wire circuits. In all other cases, the equipment must be adequate for the load involved, but never less than 60 amps. See the National Electrical Code (NEC) Section 230-79.

To determine the maximum wattage, or volt-amperes (VA), available with any size switch or breaker, multiply the amperage rating by the voltage (240). For example, a 60-amp switch or breaker will make available 14,400 VA ($60 \times 240 = 14,400$), and a 100-amp size will make available 24,000 VA ($100 \times 240 = 24,000$).

What size service?   To determine the size of service you need, begin by computing the total probable load in volt-amperes (net computed VA).

CALCULATING TOTAL PROBABLE LOAD IN VOLT-AMPERES	
Area of house* _____ square feet x 3 VA .... _____ VA	
Two appliance circuits .... 3,000 VA†	
One laundry circuit .... 1,500 VA	
Range, if used .... _____ watts‡	
Water heater, if used .... _____ VA	
Other permanently connected appliances .... _____ VA	
Total .... _____ VA	

*As calculated in Chapter 5 on page 45.

†Each of these circuits will provide 2,400 VA (20 x 120 = 2,400), but for the purpose of this calculation use 1,500 VA per circuit (3,000 VA altogether) because it is not likely that both circuits will be loaded to capacity at the same time.

‡Per the NEC, allow 8,000 watts if the rating of your range (or countertop unit plus separate oven) is 12,000 watts or less. If rated over 12,000 watts, add 400 watts for each additional kilowatt or fraction of a kilowatt. For ranges, watts and VA are considered equivalent. For examples and further details, see the new Appendix D in the 1999 NEC.

A 100-amp service is the smallest permitted by the NEC. The 100-amp service provides a maximum capacity of 24,000 VA. If the total of the "net computed VA" that you calculated on the previous page comes to more than 24,000 VA, does it follow that a service larger than 100-amp is necessarily required? No, because all the lights and appliances listed in the tabulation will not be in use at the same time. But remember that the assortment of appliances you include in your first tabulation represents your present ideas. As time goes on, you will add more appliances. A 150-amp or 200-amp service will meet future needs more adequately. In some areas, local codes are already requiring services larger than 100 amps. The larger services cost relatively little more than the 100-amp size.

Service entrance wire size   Since 100-amp service is the minimum for homes, select a wire with ampacity of 100 or more amperes from Table 4-1. For installations other than homes, NEC Section 230-42 requires that the conductors be sized to carry the load as calculated in Article 220. Examples of dwelling service calculations are found in NEC Appendix D, Examples D1a, D1b, D2a, D2b, and D2c.

DISCONNECTING MEANS

At the service location you must provide equipment, called a panelboard or load center, that allows you to disconnect all the wiring from the power source. You must also provide either circuit breakers or fuses to protect the installation as a whole as well as each branch circuit individually. The service disconnect may be either inside or outside. In conjunction with the discussion that follows, be sure to review Chapter 5: *Circuit Protection and Planning.*

Advantages of circuit breakers   Most installations today use circuit breakers because of their many advantages over fuses. If a breaker trips, you can restore service by merely flipping its handle (after correcting the problem that created the overload that tripped the breaker—see pages 193–194). You can also disconnect an individual circuit by using the breaker as you would a switch. The breakers will carry nondangerous temporary overloads that would blow ordinary fuses.

Selecting breaker equipment   Most service entrance equipment as purchased will contain one large (100-amp or larger) two-pole main breaker that will protect the entire installation and will let you disconnect the entire installation from the power source by merely turning this one breaker to the OFF position.

But there must be more breakers to protect individual circuits. You will need a single-pole breaker for each 120-volt circuit and a two-pole breaker for each 240-volt circuit. These branch-circuit breakers are not part of the equipment as purchased, but the cabinet contains an arrangement of busbars into which you can plug as many individual single-pole and two-pole breakers as there is room for. A two-pole breaker occupies the same space as two single-pole breakers. Each panelboard accommodates only as many breakers as its UL listing permits. Typically, a 100-amp panelboard will accommodate 16 plug-in breakers; 125-amp—24 breakers; 150-amp—30 breakers; and 200-amp—40 breakers. The overall rating of the

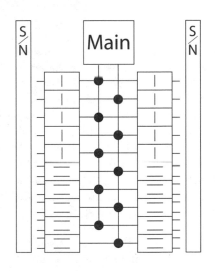

Fig. 8-1 On the left, a circuit breaker cabinet with main breaker *(Square D Company).* On the right, diagram showing space for 20 circuits. *(Square D Company)* Arrangements for mounting plug-in circuit breakers vary by manufacturer; a label in the panelboard lists catalog numbers of breakers permitted to be installed.

panelboard must equal or exceed the load served as calculated by NEC Article 220. Figure 8-1 shows a circuit breaker cabinet with its internal wiring diagram. The diagram shows space for 20 circuits. Select equipment that provides for the number of branch circuits you need plus a few additional spaces for future expansion. For help in calculating your circuit needs see the "What size service?" discussion earlier in this chapter and Table 5-3.

Split-bus breaker equipment Somewhat less popular (and in a few localities prohibited by local ordinance) are "split bus" service entrance panelboards. The NEC permits two *main* disconnects in one service panelboard. To disconnect everything from the power source, it is necessary to turn *both* of them to the OFF position. In addition to the main breakers, the cabinet may contain as many breakers as you wish (up to the

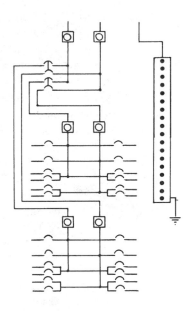

Fig. 8-2 Diagram for a split-bus circuit breaker cabinet.

capacity of the panelboard), single-pole or two-pole, each protecting its own circuit, but all of these smaller breakers must be protected by one of the main breakers. See wiring diagram of Fig. 8–2. An alternate arrangement would be for one of the main breakers to protect all the branch circuits in the service panelboard, and the other main breaker to supply a feeder to a panelboard located elsewhere.

Outdoor cabinet   In a few areas, it is common practice to use an outdoor cabinet containing the meter socket and the service disconnects. Fig. 8–3 shows a meter socket and single main disconnect in one cabinet. In mild climates, all the branch circuits could also originate in this enclosure.

Fig. 8–3 A combination meter socket plus circuit breaker in the same cabinet is very popular in some localities. *(SquareD Company)*

Selecting fused equipment   The type widely used in residential work has two main fuses mounted on a pullout block (Fig. 8–4). With the pullout in your hand, insert the fuses into their clips. One side of each fuse clip has long prongs; the entire pullout has four such prongs. The equipment in the cabinet has four narrow open slots, but no exposed live parts; the live parts are behind the insulation. Plug the pullout with its fuses into these slots; the four prongs on the pullout make contact with the live parts, completing the circuit. Plugging the pullout into its holder is the same as closing a switch with hinged blades; removing it is the same as opening a switch. You can insert the pullout upside-down to leave the power turned off.

Such fused service equipment usually contains one main pullout with large fuses protecting the entire load, plus additional pullouts to protect 240-volt circuits with

Fig. 8–4 In fused service equipment, cartridge fuses are usually installed on pull-out blocks. Replace fuses while the block is in your hand.

Fig. 8–5 A 100-amp switch. Besides main fuses, it has plug fuses for twelve 120-volt circuits, and cartridge fuses for two 240-volt circuits.

large loads such as range, clothes dryer, and water heater. There are also as many fuseholders for *plug* fuses as are needed to protect individual branch circuits: one for each 120-volt circuit, two for each 240-volt circuit. A service panelboard of this type is shown in Fig. 8–5. The wiring diagram is the same as in Fig. 8–1 except that fuses are used in place of breakers. The installation should provide for the number of circuits needed for present uses, plus a few additional spaces for future expansion.

Fuseholders For fuses of the cartridge type, the "60 amp" holder accepts fuses rated from 35 amps to 60 amps. The "100 amp" holder accepts fuses rated from 70 amps to 100 amps. The "200 amp" holder accepts fuses rated from 110 amps to 200 amps. Holders for plug fuses accept fuses up to 30 amps (the top rating for plug fuses). In new work, Type S adapters must be used; these allow installation of only the proper fuse to match the ampacity of the wires.

Split-bus fused equipment As in the case of breaker equipment, the fused equipment *may* contain two *main* pullouts, connected internally so the service wires run directly to each. To disconnect the load completely you must pull out both the main pullouts. The branch-circuit fuses must be protected by one of the main fused pullouts. The wiring diagram of connections within the equipment is that of Fig. 8–2 except that fuses replace the breakers.

Old-style fused service equipment In very old houses you will find service equipment that consists of a main switch with two hinged blades and an external handle used to turn the entire load in the building on and off. The switch cabinet also contains two *main* fuses. Usually it also contains as many smaller fuses as needed to protect all the individual branch circuits in the house. Sometimes these fuses are in a separate cabinet.

INSTALLING THE SERVICE ENTRANCE

So far we have looked at the major parts that make up the service entrance, and how to select the right kind and size of equipment. The rest of this chapter concerns the installation of these parts. Fig. 8–6 shows the details of a typical service entrance.

You will have to install insulators on the outside of the building. The power supplier will run its wires up to that point and connect them there to the service-entrance wires that you provide. You can use special cable for that purpose or run wires through conduit. Your wires will go up to the service head, down to the meter socket, and then into the house to the service equipment.

After installing the service equipment, you must properly ground the installation. Then you will be ready to wire the branch circuits.

Service equipment location The NEC requires this equipment to be located as close as practical to the point where the wires enter the house. In other words, the service wires must not run 10 or 20 feet inside the house before reaching this equipment. Decide where to put your equipment cabinet before deciding where the wires are to enter the house.

Fig. 8-6 A cross-sectional view of a typical service entrance.

The service head (see Fig. 8–18) should be mounted higher than the insulators. Let 24 to 36 inches of the service-entrance wire extend out of the head. Your power supplier will connect the service-drop wires to the service-entrance wires.

Service insulators (see Figs. 8–8 or 8–9) must be solidly mounted. Install as high as practical, but lower than service head.

Anchor cable to wall with straps (see Fig. 8–16). If you use conduit, use ordinary pipe straps.

Conduit or cable (see Figs. 8–12 and 8–13).

If you use cable, make a weathertight connection where it enters the meter socket, using outdoor type of connector (see Fig. 8–14).

Weatherproof meter and socket.

If you use cable, provide a sill plate (Fig. 8–17) where cable enters building.

Service equipment, which may be any of the types described in this chapter. Be sure to install enough branch circuits, including spares for future use. If the service equipment is located in the basement, the cable or conduit entry into the building will move down to a point opposite the equipment.

Ground wire. See page 174 in Chapter 17 for farm yard pole service grounding.

Rod may be inside or outside the wall, but should reach permanently moist earth.

Service drop

Branch circuits

Range outlet

Ground clamps

Water pipe

Ground rod

INSTALLING SERVICE EQUIPMENT AND WIRING

The heaviest loads in a house—the loads that consume the largest amperages—are located in the kitchen and basement or other laundry area. Therefore, locate the service equipment so that the circuits using large wires (to range, clothes dryer, and similar equipment) will be as short as possible. In houses with basements, a point more or less under the kitchen is desirable if the service wires can be brought into that location directly from outside. Some prefer to locate the equipment in the kitchen where it is convenient for replacing fuses or resetting breakers.

Meter and socket location The outdoor type of meter is standard today. The weatherproof meter provided by the power supplier is plugged into a weatherproof socket (see Fig. 8–7). The socket is usually, but not always, provided by the power supplier but installed by the contractor, usually about 5 feet above the ground. In the case of indoor equipment, the meter socket is installed on a substantial board near the service equipment. For a two-family dwelling, two meter sockets are installed in one sheet steel enclosure. Below the meters each occupancy will have its own service disconnect.

Fig. 8–7 Typical outdoor weatherproof kilowatthour meter and its socket.

Insulators The incoming service wires are anchored on service insulators installed on the building. They should be installed a foot or so below the highest practical point so the wires to the inside of the house, where they emerge from the service head on the conduit or cable, will slope downward toward the insulators to prevent water from following the service wires into the service head(see Fig. 8–6). Fig. 8–8 shows the type of racks used in most localities, usually with 8-inch spacing between insulators (6-inch spacing is allowed in some localities). Individual screw-point insulators (Fig. 8–9) are often used instead of multiple racks.

Fig. 8–8 Insulator rack for supporting three wires. Be sure to anchor rigidly to building.

Fig. 8–9 Screw-point insulator.

Service wire clearance   NEC Section 230-24(b) requires that cabled service wires not over 150 volts to ground (or drip loops as illustrated in Fig. 8–19) be kept 10 feet above finished grade at the point of attachment to the building; 12 feet above residential property and driveways and commercial property not subject to truck traffic; 18 feet over parking areas and driveways other than residential, and farm properties that are subject to truck traffic. No wires may come closer than 3 feet to an openable window (there is no restriction to wires above the top of window), or 3 feet horizontally and 10 feet vertically from any door, porch, fire escape or similar location from which the wires might be touched. If they pass over all or part of a roof that has a rise of 4 inches or more per foot, they must be kept at least 36 inches from the nearest point. If the roof is flatter (less than 4 inches of rise per foot), they must be kept at least 8 feet from the nearest part of the roof.

Masts   In wiring a rambler or ranch-style house it is difficult to maintain required clearances if the insulators are mounted directly on the side of the house, so a mast is used. Several types are on the market, and Fig. 8–10 shows a typical construction. The service conduit extends upward through the roof overhang and becomes the support for insulators of the type shown in Fig. 8–11. Unless your power supplier has other requirements, the service conduit should not be smaller than 2-inch rigid metal conduit. Smaller conduit could be used if braced or guyed for extra strain support, or either smaller conduit or service entrance cable could be attached to a 4 × 4-inch timber. If the roof overhang does not exceed 48 inches, the service wires may be as low as 18 inches above the overhang.

Fig. 8–10 On rambler or ranch-house construction, a mast gives the service wires necessary clearance above ground.

Fig. 8–11 Insulator that clamps to the pipe support.

Service-entrance wires   Service-entrance wires begin at the insulators where the power supplier's wires end, and they stop at the service equipment where the individual circuit wiring begins. You may use service wire cable manufactured for the purpose, or separate wires in conduit.

Service-entrance cable   Typically, service-entrance cable (Fig. 8–12) is used to bring wires into the building. One of the wires is not insulated and consists of a number of fine wires wrapped around the insulated wires. In use, the small bare wires are twisted together to make one larger wire (Fig. 8–13). The bare wire may be used only for the grounded neutral. Over all is a fabric braid or outer protective layer. This usually has a painted finish which may be painted to match the building.

CONDUCTOR | BRAID | BARE CONDUCTOR | DOUBLE RUBBER FACED TAPE | BRAID | FINISH
THERMOSET OR THERMO-PLASTIC INSULATION

Fig. 8–12 In service entrance cable the neutral wire is not insulated, but is bare and wrapped spirally around the insulated wires. The drawing shows 3-wire cable.

Fig. 8–13 When using the cable, the separate strands of the neutral bare wire are gathered into a bunch making one large wire.

STRANDS OF BARE WIRE, TWISTED

Use cable with No. 2 wires with a 100-amp main breaker or switch. Use cable with No. 3/0 wires for a 200-amp breaker or switch. As explained on page 73, in houses 30-amp and 60-amp service is not permitted but may be used in other buildings such as a free-standing studio or workshop on residential property. Use No. 8 wire with a 30-amp breaker or switch, or No. 6 with a 60-amp breaker or switch.

If the installation served by the service equipment has no significant loads operating at 240 volts, all wires in the cable, including the neutral, must be the same size. But if the installation has an electric range or water heater or other 240-volt loads consuming about one-third of the total watts, then you may use cable with the neutral one size smaller than the insulated wires—for example, No. 2 insulated wires with a No. 4 bare neutral.

Wherever the cable enters a service equipment cabinet or a meter socket, it must be securely anchored with a connector. Outdoors, a weatherproof connector must be used; a common type is shown in Fig. 8–14. The connector consists of a body and a heavy block of rubber, and a clamping nut or cover that compresses the rubber against the cable, making a watertight joint. Indoors, a less expensive connector of the type shown in Fig. 8–15 may be used.

Fig. 8–14 A watertight connector for outdoor use. To prevent leaks, apply duct seal around the cable entry.

Fig. 8–15 An ordinary connector for indoor use.

Cable is anchored to the building with straps like the one in Fig. 8–16. At the point where the cable enters the building, you must take steps to prevent rain from following the cable into the building. Duct seal (a waterproof puttylike compound) is suitable; pack it tightly into all openings around the cable. A sill plate (Fig. 8–17) is convenient but not required by the NEC. Waterproofing compound is usually supplied with the plate.

Fig. 8–16 Entrance cable must be anchored to the building every 30 inches. Use straps of the type shown. *(The Halex Company)*

Fig. 8–17 Use a sill plate packed with sealing compound where cable enters building. It keeps water out.

When you install service entrance cable, cut a length long enough to reach from the meter socket to a point at least a foot above the topmost insulator, plus another two or three feet. On this last additional length, remove the outer braid over the spirally wrapped neutral wire. Unwind these bare wires from around the cable and twist them into a single wire; see Fig. 8–13. Then install a service head of the general type shown in Fig. 8–18, letting the individual wires project through separate holes in the insulating block in the service head. The service head is designed to prevent

water from entering the top of the cable. Anchor the service head on the building about 12 to 18 inches above the topmost insulator so that, after the connection has been made to the power supplier's wires, rain will tend to flow away from the service head rather than into it.

Fig. 8–18 This service head is for cable. It prevents water from entering the cut end of the cable. A service head for conduit is supported by the conduit. *(The Halex Company)*

Fig. 8–19 If the service head cannot be located higher than the insulators, provide drip loops. Splice at bottom of loop, and insulate.

Fig. 8–20 How wires are connected to the meter base or socket. The bare uninsulated wire is always connected to the neutral center contact of the socket.

If it is impossible to locate the service head above the insulators, be sure to provide drip loops (see Fig 8–19) to keep rain out of the cable. Your power supplier will connect the ends of the wires in the cable to the service-drop wires.

Anchor the bottom end of the cable to the meter socket, and connect the three wires to the terminals in the socket, as shown in Fig. 8–20, which shows at the bottom the next length of cable running to the switch inside of the building. The cable must be supported on the building within 12 inches of the service head and meter socket, and additionally every 30 inches, using straps like those shown in Fig. 8–16.

Separate wires in conduit Instead of service entrance cable, separate Type TW, THW, or RHW wires inside conduit may be used. The neutral wire is always white; in some localities bare uninsulated wire is used. The other two wires are black, or black and red. To determine the size conduit required, see Tables 1, 4, and 5 in Chapter 9 of the NEC.

Conduit is available in three types: rigid, intermediate, and electrical metallic tubing, all described in Chapter 12. Cut pieces as long as required and prepare them in the manner described in Chapter 12 depending on the type used. At the top of the conduit install a service head of the type shown on the right in Fig. 8–21. At the point where the conduit is to enter the house, use an entrance ell such as shown on the left in Fig. 8–21. This has a removable cover which makes it easy to help the wires around the sharp bend while pulling them into the conduit.

Fig. 8–21 Service head *(at right)* is for use when wires run through conduit. Use with special entrance ell *(at left)* which has a removable cover. *(Hubbell Electrical Products)*

After the conduit itself is completely installed, pull the wires into it. For the relatively short lengths involved, the wires can usually be pushed in at the top and down to the meter socket; from there, other lengths are pushed through to the inside of the house. If the conduit is long or has bends in it, use fish tape to pull the wires through. See Fig. 12–8 and related text for how to use fish tape. A length of galvanized clothes line or similar wire will serve the purpose for short lengths.

Concentric knockouts The cabinets of service switches or circuit breakers are too small to permit knockouts of all the sizes that might be required in all circumstances, so concentric knockouts are provided (Fig. 8–22). Remove the center section if you need the smallest size; remove the two smallest sections for the next size, and so on. Be very careful to remove only as many sections of this intricate knockout as required for your particular installation.

Fig. 8–22 Concentric knockouts are convenient, but great care must be used in removing each section.

MAKING GROUND CONNECTIONS

The service switch or circuit breaker has two heavy terminals to which the two incoming hot wires are connected. It also has a neutral busbar on which you will find several large connectors and a number of smaller terminals. Connect the incoming white or bare neutral to one of the large connectors, and the ground wire to the other. The smaller terminals are for the white wires of the individual circuits or feeders. If there are enough terminals, bare grounding wires of circuits can also be connected to the neutral bar *in service equipment if acceptable to the inspector*. If additional terminals are required, an accessory equipment grounding bar must be used. The enclosure will have mounting means for this ground bar.

Solid neutral   Whether your service equipment contains breakers or fuses, the incoming neutral wire must never be interrupted by a breaker or fuse. Therefore in a three-wire 120/240-volt installation, while there are three service-entrance wires, the main breaker is only a two-pole, and in fused equipment there are only two main fuses. The neutral is call a "solid neutral"—a neutral not protected by breaker or fuse. Such equipment then is called "three-wire SN."

Grounding   Every installation of the types discussed in this book must be grounded. The technical reasons for grounding are discussed in Chapter 7. A good, carefully installed ground is absolutely necessary for a safe installation. Grounding is accomplished by running a ground wire from the incoming neutral service wire (from a point where it is connected to the neutral busbar in the service equipment) either to a grounding electrode system or, if not available, to a grounding electrode.

Grounding electrode system   All of the following which exist at the structure must be bonded together to form a grounding electrode system:

- 10 feet or more of buried metal water pipe
- the effectively grounded metal frame of the building
- a concrete-encased electrode consisting of 20 feet or more of ½-inch reinforcing steel or No. 4 copper wire encased by at least 2 inches of concrete and located near the bottom of a concrete foundation or footing that is in direct contact with the earth
- a ground ring of No. 2 or larger bare copper wire encircling the building

If only one of these exists it may serve as the grounding electrode, *except* for water piping which *must* be supplemented by at least one other electrode. If none of these is available, use a "made" electrode.

Made electrodes   If there is no underground water system, a good ground becomes a problem. If none of the above options for a grounding electrode system is available, the NEC allows for several alternatives (called "made electrodes") that may serve as the grounding electrode:

- underground metal piping (except gas) or tanks
- 8-foot driven ½-inch copper or ⅝-inch steel rod or ¾-inch galvanized steel pipe
- a 2-square-foot metal plate buried at least 30 inches below the surface

Ground rod installation is discussed on pages 178–179. If a ground rod alone is used, the ground wire never needs to be larger than No. 6 regardless of the size of the wires in your service equipment.

Bonding the busbar   Service equipment cabinets contain a copper neutral busbar on which are installed several large solderless connectors for the heavy incoming neutral wire and the ground wire, plus as many smaller connectors or terminal screws as needed, one for the grounded wire of each 120-volt circuit. As purchased, the neutral busbar may be insulated from the cabinet. If you use this equipment as service equipment you *must* bond the neutral busbar to the cabinet. It must *not* be bonded if the breaker or switch is used, for example, as a disconnecting means for an appliance or other load, or is used as a panelboard on the load side of a service. If you have to bond the busbar to the cabinet, it is a simple matter. In some brands of equipment the neutral busbar is provided with a green bonding screw that you must tighten securely. In other brands, the cabinet will contain a flexible metal strap that is already bonded to the cabinet, and you must connect its free end to one of the connectors or screws on the neutral busbar. The equipment grounding busbar will always be bonded to the enclosure, whether within the service equipment enclosure or at a remote panelboard.

Installing the ground wire There is never more than one ground wire from the service to the grounding electrode(s). There is no objection to using insulated wire, but bare wire costs less and is usually used for the ground wire. Keep the ground wire as short as possible. All grounding connections to the water pipe must be made within the first five feet of where the water pipe enters the building.

Size The minimum ground wire size is based on the size of your service wiring. Larger size ground wire is optional. With No. 2 or smaller service wires, No. 8 is permitted for the ground wire but it must be enclosed in armor or conduit. It is more economical and practical to use No. 6 because it doesn't have to be protected.

With No. 1 or No. 1/0 service wires, No. 6 is the minimum ground wire size. With No. 2/0 or 3/0 service wires, No. 4 is the minimum ground wire size.

Armor or conduit is not required for No. 6 or No. 4 ground wire. Either size must be fastened with staples to the surface over which it runs. It must be installed in a way that minimizes danger of physical damage after installation. Physical protection for the ground wire can also be provided by rigid nonmetallic (PVC) conduit.

Connecting At the service equipment, connect the ground wire to the neutral busbar to which you have connected the neutral of the service. Run it to the grounding electrode(s), and there connect it to a ground clamp of the general style shown in Fig. 8–23. This clamp must be made of galvanized iron if installed on an iron pipe, but copper or bronze if it runs to a copper pipe or rod.

Fig. 8–23 Typical ground clamp for connecting ground wire to water pipe. *(The Halex Company)*

Bonding at the service entrance A careful job of bonding is essential to assure effective grounding.

Service-entrance wires If your service-entrance wires are in conduit, follow A of Fig. 8–24. Install a grounding bushing (Fig. 8–25) on the end of the conduit inside the cabinet and run a jumper (bare or insulated, of the same size as the ground wire) from the bushing to the grounding busbar. The screw of the bushing bites down into the metal of the cabinet, contributing toward a good continuous ground and preventing the bushing from turning and loosening which would impair the continuity of the ground. If your service uses service-entrance cable with its bare neutral wrapped around the insulated wires, as in Fig. 8–12, the grounding bushing is not required; follow B of Fig. 8–24.

Fig. 8–24 Typical grounding connections. Only the white wire is shown. The wire marked "bonding jumper" is usually part of the service equipment as purchased; it could be either a green screw or a flexible strap.

Fig. 8–25 Grounding bushing.

Fig. 8–26 Two-piece threaded hub. Serrations on both pieces bite into enclosure metal, making a bonding jumper unnecessary. O-ring in outside piece forms a raintight seal.

When you use a threaded hub bolted to the enclosure, or a hub of the two-piece type shown in Fig. 8–26, bonding of the conduit to the enclosure is automatic and the grounding bushing and jumper are not necessary.

Ground wire   If you use armored ground wire, bond the connector on the cable to the cabinet, and at the electrode use a ground clamp similar to the type shown in Fig. 8–23; connect the wire to the terminal screw, and clamp the armor in the separate clamping device of the ground clamp.

If the ground wire is protected by conduit, bond the conduit to the cabinet with a locknut and grounding bushing. A jumper from that bushing to the neutral busbar is required, as shown for the service conduit in A of Fig. 8–24. Use a ground clamp into which you can thread the conduit, and connect the ground wire to a terminal screw of the clamp.

If the ground wire is protected by rigid nonmetallic (PVC) conduit, bonding of the raceway is not necessary. *Caution:* When using either armor or conduit to protect the ground wire, great care must be taken to securely and permanently bond the armor or conduit at both ends—to the cabinet and to the ground clamp at the electrode. Unless this is done, the resulting ground will be much less effective than when using unprotected wire.

Fig. 8–27 Install a jumper around the water meter.

Water meter   A bonding jumper must be installed around the water meter as shown in Fig. 8–27. Use two ground clamps and a length of wire of the size used for the ground. This prevents the ground from being made ineffective if the water meter is removed. Some water meters also have insulating joints.

EMERGENCY BACKUP GENERATORS

There may be times when the main power supply is cut off due to a storm or other factors. If you live far from town, or in a lightning-prone area, your electric service may be subject to occasional or frequent outages. Experience will tell you whether a generator would be a good investment for you. Choice of a portable or permanently installed type will depend on your particular needs. Fuel types include gasoline, natural gas, diesel fuel, or liquid propane. Be sure to check local requirements for storage of fuel and the noise level permitted.

Portable or permanently installed?   Small portable generators can be used to power a few individual appliances such as the refrigerator, freezer, and a lamp or two. Be sure the extension cords you use to connect the appliances to the generator are sized to carry the specific loads. Check the watts/volts ratings on the nameplates of the appliances and on the extension cords. For convenience you may want a permanently installed generator. There are systems that automatically start the generator when the normal power source fails, and automatically transfer the load(s), but for the usual residential situation a system for manually starting the generator and transferring the load(s) will usually suffice.

Choosing the correct size   Generators are rated and priced according to the wattage they produce. Noise level and rate of fuel consumption (running time) are other factors to consider in making a selection. The size you need depends on whether you want to have the load of the entire house or farm on during an outage, or only a few selected circuits. A limited amount of rewiring may be required in order to separate the circuits for your essential equipment from the rest of the load, or in the case of a farm, the selected loads from the total load. If in doubt about how to proceed, consult with a professional.

See pages 44 and 169–171 for information on the calculation of residential and farm loads. For motors, convert the horsepower to watts by multiplying by 746. Use a generator with 25 percent capacity above the proposed connected load to permit motors to start. If your load includes hard-to-start motors, such as a well pump, it would be prudent to size the generator capacity to as much as three times the connected load in order to permit the motor to start. If some of the load is 240-volt, be sure the generator you choose will supply 240 volts, and not 120 volts only. A generator that is automatically started must be sized large enough to carry the entire load connected to it, startup as well as continuing. With a manually started generator you can turn off all equipment before starting the generator, and then restart selected equipment individually to minimize the load.

Isolate generator from main power supply   It is absolutely necessary that it not be possible for your generator and the incoming lines from the power supplier to be connected together. The principal reason for making sure your generator is isolated from the power supplier's lines is to protect the workers who are out on the power supplier's system trying to restore power—they could receive a serious shock or be electrocuted by current coming from your generator. The NEC requires that a sign be placed at the service location advising of the presence of the standby source.

Another reason for keeping your generator separate is to avoid damage to it and your wiring and appliances. Sophisticated equipment would be required for your generator and the power supplier's system to work together because they would have to match exactly in voltage and frequency.

The required separation of generator and power supplier's lines can be accomplished with a properly installed transfer switch that meets the requirements of the NEC. It is recommended that you check with a professional or with your power supplier for information about such transfer equipment.

Grounding the generator   The generator frame must be grounded to the same electrode as that used for the service. If the generator is not permanently installed, but moved into position when needed, be sure there is a reliable means to accomplish this grounding, such as a flexible cable and a ground clamp, whenever the generator is in use. Be sure that after the transfer to the generator the identified (white, neutral) wire is still bonded to the grounding electrode.

CHAPTER 9

Outlet and Switch Boxes

IN ORDER TO LIMIT SHOCK and fire hazard from poorly made or deteriorated connections and splices in wires, and to protect the wiring from physical damage, the National Electrical Code (NEC) requires that every switch, every outlet, and every joint in wire or cable must be enclosed in a box, and every fixture must be mounted on a box (with the exception of some fluorescent fixtures described on pages 154–155). Keep in mind that all explanations in Chapters 9 through 15, except Chapter 13, are for "new work"—the wiring of a building while it is being built. The basic principles for "old work"—the wiring of a building after it is built—are covered in Chapter 13. Study new work thoroughly in order to better understand old work.

PLANNING YOUR BOX INSTALLATION

Many boxes are made of metal with a galvanized finish. Boxes made of insulating materials such as PVC, fiberglass, and other nonmetallic material are also common. Such boxes are used mainly with nonmetallic-sheathed cable and are discussed on page 167. Metal boxes and their accessories are the focus of this chapter.

Types of boxes   Boxes are called switch boxes or outlet boxes depending on their shape and purpose. Wires and cables must be brought into the boxes through openings of the appropriate size, and "knockouts" are provided for this purpose. These are sections of partially punched out metal that are easily removed to form openings by loosening the knockout with a sharp blow on a screwdriver placed against it and then removing with pliers.

Switch boxes   The most common type of switch box is shown in Fig. 9–1. The mounting brackets on the ends of each box can be adjusted to bring the front of the box flush with the surface of the wall. Two "ears" with tapped holes for screws hold the switches, receptacles, and similar devices installed in the boxes. Switch boxes are sometimes used for fixtures of the wall-bracket type weighing up to 6 pounds, with a box located at each point where a bracket is to be installed.

The NEC requires boxes to have a depth of at least $^{15}/_{16}$-inch where a device such as a switch or receptacle is to be installed. Most are at least 1½ inches deep, but the

Fig. 9–1 A typical switch box used to house switches, receptacles, etc. The sides are removable.

Fig. 9–2 Two single boxes are easily changed into one larger "two gang" box. Still larger boxes of three or four or more gangs are made the same way. Ready-made two-gang and larger boxes are also available.

deeper ones (2 to 3½ inches deep) are handier and more generally used.

Each box holds one device. When two devices are to be mounted side by side, two boxes can be changed into a single "two gang" box by simply throwing away one side of each box and bolting the boxes together as in Fig. 9–2. Use the screws that are part of the boxes.

Outlet boxes Fig. 9–3 shows the most common outlet box. It is octagonal in shape and available in three sizes: 3¼-inch, 3½-inch, and 4-inch. The 4-inch size permits more wires to be used, avoids cramping, and in general reduces the time required for installation. If space permits, use a box at least 1½ inches deep.

Outlet boxes must always be covered. A great many different covers are available. Some common ones are shown in Fig. 9–4. The cover at A is a blank cover for a box that is used only to hold a splice or tap. The one at B is a drop cord cover—the opening for the drop cord is fitted with a smooth bushing to eliminate sharp edges. At C is a spider cover used to mount surface-type switches. At D is a cover with a duplex receptacle. At E is a keyless receptacle for a lamp. Another lamp receptacle is shown at F, this one with a pull chain.

Fig. 9–3 The common octagonal box.

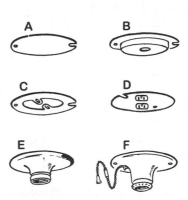

Fig. 9–4 Every outlet box must be covered. Some common covers: A) blank cover; B) drop cord cover; C) spider cover; D) cover with duplex receptacle; E) keyless receptacle for lamp; F) same as E but with pull chain.

Fig. 9–5 This box is known as a "handy box" or "utility box." Use it for permanently exposed surface wiring as in basements, garages, etc. The box and covers have rounded corners for safety. *(Hubbell Electrical Products)*

Fig. 9–6 Four-inch square boxes and covers. Use covers on the left and middle when wiring is concealed. For permanently exposed wiring, use the type of cover shown on the right. *(Hubbell Electrical Products)*

Handy boxes   If an ordinary switch box is mounted on the surface of a wall, as in a basement or garage, the sharp corners of the box and cover are a nuisance. In such locations use a box with rounded corners, known as a "handy box" or "utility box" as shown in Fig. 9–5.

Square boxes   Boxes 4 inches square are large enough to hold two devices. Fig. 9–6 shows a square box and the special covers used with it. These boxes are used for surface mounting as well as for general-purpose wiring and are especially handy when many wires must enter the same box.

How to calculate correct box size   NEC Section 370-16(a) specifies the maximum number of wires of the same size permitted in a metal box. Do not crowd a box to its limit. Doing so makes the work more difficult and time-consuming, and it runs a risk of shorts and grounds. Table 9–1 shows the interior cubic inch capacities of the more common standard metal boxes as taken from the complete list found in NEC Table 370-16(a). Nonstandard and nonmetallic boxes have their cubic inch capacities marked on them.

The total volume of an enclosure includes plaster rings, domed covers, extension rings, etc., which are marked with their volume in cubic inches. The space required for each conductor is shown in Table 9–2.

To find the minimum size box that is permitted, add up the required volumes listed in the box at the top of next page. Choose all volumes from Table 9–2. Then match the total calculated volume in cubic inches to a box size in Table 9–1.

To calculate minimum box size, add up these required volumes, choosing volumes for wires from Table 9–2 below and then matching total calculated volume to a box size in Table 9–1.

1 VOLUME .... for each wire originating outside the box and terminating or spliced within the box.

1 VOLUME .... for each wire passing through the box without splice or termination (as may happen in conduit wiring).

1 VOLUME .... for each fixture wire. *Exception:* four or fewer fixture wires and one equipment grounding wire entering the box from a domed fixture canopy are not counted.

1 VOLUME .... for any number of internal cable clamps (Fig. 9–14), basing the volume on the largest wire in the box. Do not add for external cable connectors (Fig. 9–11).

1 VOLUME .... for any number of fixture studs (Fig. 9–15 top), basing the volume on the largest wire.

1 VOLUME .... for any number of hickeys (Fig. 9–15 bottom), basing the volume on the largest wire.

2 VOLUMES .... for each yoke or strap containing switches, receptacles, pilot lights, etc., basing the volume on largest size wire connected to a device on that yoke, and then doubling it for two volumes.

1 VOLUME .... for any number of equipment grounding conductors, basing the volume on the largest equipment grounding conductor in the box.

Table 9–1 INTERIOR VOLUME OF STANDARD BOXES

KIND OF BOX	SIZE IN INCHES	INTERIOR VOLUME IN CUBIC INCHES
Outlet box	4 x 1¼ ⎫ ROUND	12.5
	4 x 1½ ⎬ OR	15.5
	4 x 2⅛ ⎭ OCTAGONAL	21.5
	4 x 1¼ SQUARE	18.0
	4 x 1½ SQUARE	21.0
	4 x 2⅛ SQUARE	30.3
	4¹¹/₁₆ x 1½ SQUARE	29.5
	4¹¹/₁₆ x 2⅛ SQUARE	42.0
Switch box	3 x 2 x 1½	7.5
	3 x 2 x 2	10.0
	3 x 2 x 2¼	10.5
	3 x 2 x 2½	12.5
	3 x 2 x 2¾	14.0
	3 x 2 x 3½	18.0
Handy box	4 x 2⅛ x 1½	10.3
	4 x 2⅛ x 1⅞	13.0
	4 x 2⅛ x 2⅛	14.5

Table 9–2 VOLUME REQUIRED PER CONDUCTOR

WIRE SIZE	VOLUME IN CUBIC INCHES PER WIRE
18	1.50
16	1.75
14	2.00
12	2.25
10	2.50
8	3.00
6	5.00

Example of box size calculation   Consider a feed-through receptacle location with two No. 12-2 with ground cables entering the box through internal cable clamps. From Table 9–2, each No. 12 wire requires 2.25 cubic inches. The grounding conductors are also No. 12, so the calculations are:

4 No. 12 circuit wires	4 x 2.25 cubic inches = 9.00 cubic inches
2 No. 12 ground wires (count as 1)	1 x 2.25 cubic inches = 2.25 cubic inches
2 internal clamps (count as 1)	1 x 2.25 cubic inches = 2.25 cubic inches
Duplex receptacle (count as 2)	2 x 2.25 cubic inches = 4.50 cubic inches
	Total required 18.00 cubic inches

Thus from Table 9–1, the smallest square outlet box permitted is a $4 \times 1\frac{1}{4}$-inch box, and the smallest switch box permitted for this application is a $3 \times 2 \times 3\frac{1}{2}$-inch box. Both are 18 cubic inches.

INSTALLING BOXES

NEC Section 210-70 requires the following areas to be equipped with at least some lighting controlled by a switch: every livable room; bathrooms, hallways, stairways, attached garages; detached garages if wired; and the following spaces if used for storage or containing equipment requiring servicing—attics, underfloor spaces, utility rooms, basements. In addition, each entrance into the house must have an outdoor light controlled by a switch inside the house. All required lighting must consist of permanently installed lighting fixtures in bathrooms and kitchens, but in other habitable rooms it may be provided by floor or table lamps plugged into receptacles controlled by a wall switch.

Boxes for switches should be located near doors so the switches can be easily found as the door is opened. Consider which way the door is to swing and install the box on the side opposite the hinges, with the center of the box 48 to 52 inches above the floor. For receptacles, the usual height is 12 to 18 inches above the floor. At least one should be installed in each room where it will always be easily accessible for a vacuum cleaner and other portable items.

Mounting switch boxes   Boxes must be firmly secured to studs. Some boxes come with nails attached. If mounting nails pass through the box, they must not be more than ¼ inch from the back or ends. It is usually faster to use boxes with mounting brackets (Fig. 9–7). Always install boxes so their front edges are flush with the surface of the finished wall.

If the box is not adjacent to a stud, use an octagonal or square box with a plaster ring and a hanger (see next paragraph).

Fig. 9–7 Boxes with mounting brackets save time.

Fig. 9–8 Hangers are adjustable in length.

Fig. 9–9 Hanger with box supported in ceiling.

Fig. 9–10 Boxes preassembled with hangers are often used.

Mounting outlet boxes If the box is to be installed near a stud, use an outlet box with a mounting bracket such as shown in Fig. 9-7. If the box must be mounted between studs or joists, use a hanger of the type shown in Fig. 9-8. It is adjustable in length and has a fixture stud sliding on it. Remove the center knockout in the box, slide the stud to the position wanted, push the stud through the knockout, and drive the locknut on the inside of the box securely home. Then nail the hanger to the studs or joists so the front edge of the box (or the cover mounted on it if it is the type shown in Fig. 9-6) will be flush with the wall or ceiling as shown in Fig. 9-9. Even handier are preassembled boxes with hangers as shown in Fig. 9-10.

Wiring at boxes Regardless of the wiring method that you use, let 6 to 10 inches of wire extend from the face of each box. The length depends on your choice of two methods of connecting the wire to the terminal screw as discussed on pages 35–36 and shown in Figs. 4-13 and 4-14. Remember that if a switch or receptacle must later be replaced, wires that seem just right in your original installation may be too short for the replacement device which might have terminal screws in a different location.

Anchoring connectors When cable of any style is used for wiring, the NEC requires that it be securely anchored to each box that it enters. There are many kinds of connectors for this purpose; an assortment is shown in Fig. 9–11. The connector at A is used for ordinary purposes, the one at B for a sharp 90° turn, and the one at C when two pieces of cable must enter the same knockout. To anchor the cable, first remove the locknut from the connector. Attach the connector to the cable with the clamp screw as in Fig. 9–12, then slip the connector through the knockout, and install the locknut inside the box as shown in Fig. 9–13. Be sure to drive the locknut down solidly so the lugs on it actually bite into the metal of the box to form a good continuous ground. Some boxes have built-in clamps that securely hold the cable entering the box, making separate connectors unnecessary. Typical boxes of this kind are shown in Fig. 9–14.

A **B** **C**

Fig. 9–11 Connectors used in anchoring cable to boxes.

Fig. 9–12 Connector is first attached to the cable by means of a clamp screw.

Fig. 9–13 Connector is then anchored to the box by means of a locknut inside the box.

Fig. 9–14 With boxes of this type, separate connectors are not used. The boxes have clamps to hold the cable.

INSTALLING SERVICE EQUIPMENT AND WIRING

Fixture studs   Small, lightweight fixtures are often mounted directly on outlet boxes, anchored by screws entering the "ears" of the box as explained at the beginning of this chapter under the heading "Switch boxes." Heavier fixtures need additional support. Install a fixture stud (see Fig. 9–15) in the back of the box using stove bolts. This is unnecessary if a hanger (Fig. 9–8) is used to support a box mounted between studs or joists. With a fixture stud, a hickey may be needed between the stud and fixture.

Fig. 9–15 Fixture stud and hickey.

Boxes for suspended ceiling (paddle) fans   Boxes for the support of ceiling fans must be Listed for the purpose. Fans weighing more than 35 pounds must be supported independently of the box. Consider using a box listed for fan support at every ceiling outlet location where a paddle fan might be installed at a later date, because installing a fan support box at an existing outlet can be difficult.

Boxes for recessed fixtures   If the bottom of your fixture will be flush with the ceiling it is called a "recessed fixture." If your recessed fixture includes a junction box marked for 60° branch circuit wiring, ordinary branch circuit wiring can be connected directly to it. If the marking on the recessed fixture calls for a high temperature wire, then mount a junction box in the attic or joist space at least 12 inches away from the fixture for connection of high temperature fixture wire to the branch circuit. See page 156 in Chapter 15 before installing.

Boxes for interchangeable devices   Up to three interchangeable devices can be installed in an ordinary single-gang box. Fig. 9–16 shows a face plate with three openings, and an assortment of devices such as switches, receptacles, and pilot lights. Plates are also available with one or two openings. The plate is supplied with a metal strap with three openings on which you can mount any combination of one, two, or three of these specially sized devices. The three devices can then be installed in a single-gang box. Three standard devices would require a three-gang box which costs more, takes more space and requires much more installation time. *Caution:* Remember that the number of wires entering any box is limited; review pages 92–93 on calculating minimum size boxes. When a switch box is not large enough for the wires necessary to connect to three devices on one strap, use a 4-inch-square box and a single-gang plaster ring as shown in Fig. 9–6.

Fig. 9–16 Interchangeable devices permit 3 switches, receptacles, and the like to be installed in a single-gang box.

SELECTING SWITCHES

While all common switches have the general appearance shown in Fig. 2–1, there are three varieties that you should understand:

Single-pole is used to turn a light on or off from one point. It has two terminals and the words ON and OFF on the handle.

Three-way is used to turn a light on or off from two separate locations. It has three terminals and a plain, unlabeled handle. For the wiring of three-way switches see pages 52–55.

Four-way is used in combination with three-way when a light must be turned on or off from more than two locations. Use three-way switches at two of the points, and four-way at the remaining points. The four-way has four terminals and a plain, unlabeled handle. For the wiring of four-way switches see pages 55–57.

Switch ratings   Switches are designed for use at an amperage and voltage not higher than the limits stamped into their metal mounting yokes. Some switches are rated "10A 125V–5A 250V"—meaning that the switch may be used to control loads not over 10 amps if the voltage is not over 125 volts, but only 5 amps if the voltage is higher but not over 250 volts. Most are rated 15A 120V, or 15A 120/277V. Other switches are available for higher amperages.

Grounding for switches   The yokes of all toggle switches, including dimmers as described below, are required to be grounded so that a metal face plate, whether installed or not, will be grounded. The switch yoke can be grounded to the grounded metal box, or an equipment grounding conductor must be connected to a green terminal screw on the yoke.

Types of switches   In addition to the common general-use switches, there are switches with a variety of special features that enhance comfort, convenience, and safety. Some of these are described here.

AC general-use switches   The most commonly used switch today is called the "ac general use" type in the NEC. It is very quiet in operation and lasts a long time. It is identified by the letters "AC" following the ampere and voltage rating on the yoke. As the name implies, it must not be used on dc circuits. On ac, it may be used for any purpose with two exceptions: a) it must not be used to control tungsten-filament lamps (ordinary lamps) at voltages above 120 volts; b) if used to control a motor, it may be used only up to 80 percent of its ampere rating.

AC-DC general-use switches   Older style switches are called "ac-dc general use" in the NEC. If the switch does not have "AC" at the end of its rating, it is an ac-dc type. This style may be used on dc-only circuits as well as on ac circuits. If your wiring was installed many years ago, the switches are likely to be the ac-dc type. If one of them fails, you may replace it with the more readily available ac-only type.

Lighted switches   Switches are available with a tiny neon light in their handles to make it easy to locate the switch in the dark. The light has an exceedingly long life, possibly 20 years, and consumes so little power that it costs perhaps one cent per year to operate.

Quiet switches   Ordinary switches can make an audible click when turned on and off. The "ac general use" switch is very quiet. Also available is a completely silent switch that uses a bit of mercury in a glass tube to make and break contact. Mercury switches will not operate unless installed in a vertical position.

Dimmer switches   Where it might be desirable to control the level of brightness, such as in dining rooms or family rooms, ordinary switches can be removed and replaced with special dimming switches. Some dimmers are designed to be used only with incandescent lamps and others are for use only with fluorescent lighting.

One inexpensive type of dimmer switch provides HIGH–OFF–LOW positions and controls up to 300 watts. It can be used to replace only an ordinary single-pole switch, not a three-way. The somewhat more expensive type controls brightness continuously from off to full bright. Some models are available in the three-way type, replacing one of a pair of three-way switches.

FACE PLATES

Receptacles and switches when installed are covered with face plates (also called wall plates) of the type shown in Fig. 9–17. These are available in molded plastic and nylon in an assortment of colors and in other materials such as steel, brass, or ceramic. Use the kind you like best. Metal face plates must be grounded.

Fig. 9–17 Every switch and receptacle must be covered with a face plate.

CHAPTER 10

Basic Wiring Procedures

IN RESIDENTIAL AND FARM wiring, three different wiring systems are in common use: nonmetallic-sheathed cable, armored cable, and metal conduit of four types—rigid galvanized conduit, intermediate metal conduit (IMC), electrical metallic tubing (EMT), and flexible conduit. Other methods not covered in this book include electrical nonmetallic tubing and rigid nonmetallic conduit. Both have excellent resistance to moisture and corrosion and are light in weight and easy to install. For description, see the National Electrical Code (NEC) Articles 331 and 347. The knob-and-tube wiring system, not permitted for new work since the 1975 NEC, is only for extensions in existing work (see page 130).

Local codes sometimes prohibit one or more of these systems. What is prohibited in one locality may be required in another. For farm buildings, nonmetallic-sheathed cable is used almost exclusively. It is also used extensively in residences and larger buildings, though in larger cities it is sometimes prohibited for new buildings. Armored cable is also widely used for non-farm wiring. Follow local custom. If you are not sure, consult your power supplier or electrical inspector.

Many wiring procedures are the same regardless of which system is used. These procedures are explained in this chapter. It is essential to understand them thoroughly. The diagrams and explanations in this chapter use cable as a model, but the principles apply generally to all wiring systems. Details specific to cable or conduit are described in Chapters 11 and 12.

PLANNING THE INSTALLATION

Before doing any actual wiring, you must make a plan. Decide upon the location of each outlet and each switch. Be generous in the number of outlets and switches. Review Chapter 2 on planning an adequate installation. Remember that adding an outlet later costs much more than including it in the original job.

Which outlets on which circuit? Your first idea may be to put all your basement outlets on one circuit, all first-floor outlets on the next circuit, all second-floor outlets on still another circuit, and so on. If you do that and a breaker trips or a fuse blows, an entire floor will be dark. Instead, put different parts of any floor on different circuits. Then at least part of each floor will still be lighted if any circuit goes out. One of the house circuits can be extended to an attached or separate garage that has small power requirements, or separate circuits can be used for a garage and accessory buildings requiring power (greenhouse, studio, machine shop, etc.). Wiring for garages and accessory buildings is discussed at the end of this chapter.

Diagram your circuits   After you have decided which outlets are to go on each circuit and where you want single-pole switches and three-way switches, draw a diagram of each circuit showing how the cable is to run from one outlet to the next, from outlet to switches, and so on. Fig. 10–1 is a typical diagram for an installation using cable. It shows eight outlets, plus three single-pole switches and a pair of three-way switches. The outlets are labeled A, B, C, D, E, F, G, and H. Switches have all been labeled S. The switch controlling outlet B is indicated as S-B, the one controlling outlet G as S-G, the one controlling outlet H as S-H, and the two 3-way switches controlling outlet F as S-F-1 and S-F-2. (Outlet A is a light controlled by a pull chain and is included here as part of the general explanation. Note that pull chains are not recommended; switches are preferred for convenience and safety.)

This diagram still does not tell you how to connect the individual cable wires so that all the parts will work properly. Draw a second diagram showing the wires separately as if they were open wires instead of in cable. In order to simplify the diagrams and explanations, the cable's bare grounding wire is omitted here, but requirements and instructions for connecting it are given on pages 115-116 in

Fig. 10–1 Before doing any wiring, make a rough layout of each circuit showing location of each outlet and how cable will run from box to box. Shown here is a layout of a circuit with 8 outlets labeled A to H, 3 single-pole switches S-B, S-G, and S-H, and a pair of 3-way switches S-F-1 and S-F-2. The wires for this circuit begin in the basement and run to main floor living room, kitchen, and dining room, and then to a bedroom.

Chapter 11 under the heading "Use cable with grounding wire." Fig. 10–2 shows the same outlets as Fig. 10–1, and each outlet has been labeled the same as in Fig. 10–1. Note that the same wire identification scheme has been used as in Chapter 6: a light line like this ——— for a white wire, a heavy line like this ——— for a black wire (or other color, but not white or green), and a heavy broken line like this – – – for the wire between a switch and the outlet it controls.

Fig. 10–2 The same outlets as in Fig. 10-1 but in greater detail, now showing how the wires inside the cable are connected.

If you refer to the diagramming of basic circuits covered in Chapter 6, you will see that Fig. 6–3 is the same as outlet A in Fig. 10–2 (if you disregard the cable that runs on to B). Fig. 10–3 shows this same outlet completely installed with an outlet box. Following the principles presented in Chapter 6, you run the white wire from SOURCE to the outlet, and you run the black wire from SOURCE to the outlet.

Fig. 10–3 Outlet A of Figs. 10–1 and 10–2, completely installed.

To better understand the wiring of each additional outlet, consider the cable that runs to it from the *previous* outlet as the source for the *new* outlet. For example, the cable that runs from outlet A to outlet B becomes the source for outlet B.

FEEDING AND CONNECTING THE WIRES

If you are using nonmetallic-sheathed or armored cable, you will need to remember the important NEC requirement described here regarding wire color in making connections at switches. Following that, the basic guidelines for making connections apply to either cable or conduit.

White wire re-identified as black wire in switch loops   In Fig. 10–4, if you disregard the cable running to C, outlet B is the same as that in Fig. 6–5, so a detailed explanation may seem unnecessary. However, according to everything you have learned up to this point, and as shown in Fig. 10–2, *both* wires from outlet B to switch S-B should be *black,* but the two-wire cable that you are going to use contains one black and one white wire. How can you comply with the NEC requirement? NEC Section 200-7 permits a white wire to be used where a black wire *should* be used—but only in a switch loop (the cable running from an outlet to a switch), and the white wire must be re-identified as black wherever it is visible and accessible. This can be accomplished by painting, wrapping with black tape, or sleeving the wire with a tube of either black plastic insulation removed from a larger wire or with plastic shrink tubing applied with a heat gun.

Fig. 10–4 Outlet B of Figs. 10–1 and 10–2. This is an important diagram. It shows how to connect white wire in cable to the switch.

S-B

Re-identify white wire as black (see text)

From A

To C

To F   C

D

E

From B

Fig. 10–5 Receptacle outlets are easy to connect, as this diagram shows.

Wiring at outlets and switches   It is easy to make the right connections if you remember that each fixture *must* have one white and one black wire connected to it, and if you observe the following simple steps:

1 At the switch, connect the two wires of the cable to the switch.

2 At the outlet, connect the white wire from SOURCE to the fixture, as usual.

3 At the outlet, connect the *black* wire from SOURCE to the *white* wire of the cable that runs to the switch, and re-identify the white wire as black. This is the only case where a white wire may be connected to a black.

4 Connect the black wire of the cable that runs to the switch to the fixture, as usual.

5 The two wires running on to the next outlet are connected to the two incoming wires (from source) in the outlet box—black to black, and white to white.

When outlet B is properly installed according to these five steps, it will be connected as shown in Fig. 10–4, which complies with NEC Section 200-7.

Outlets C, D, and E are wired as shown in Fig. 10–5. Receptacles have double terminal screws so that the two wires from two different pieces of cable can easily be attached as shown. In the case of C, it is necessary to connect three different wires to each side of the receptacle, but there are only two terminal screws. *NEC Section 110-14(a) prohibits more than one wire under one terminal screw.* Splice all the blacks together and all the whites together, adding a short piece of wire called a "pigtail" to the white set and another to the black set, and connect these pigtails to the receptacle terminal screws. See Fig. 4–15.

Outlet F in Fig. 10–2, if you disregard the cable running on to the next outlet G, is the same as Fig. 6–14 in Chapter 6. Wire it as shown in Fig. 10–6. Run two-wire cable from the outlet to the first three-way switch S-F-1 and three-wire cable from

Fig. 10–6 The wiring of outlet F of Figs. 10–1 and 10–2, with the two 3-way switches that control the outlet. This is the diagram for any outlet controlled by 3-way switches when using cable. Study carefully the colors of the wires in this cable. Compare it with Fig. 6–14.

there to the second three-way switch S-F-2. Again you meet the problem of the proper colors of wire. Remembering the steps outlined in connection with outlet B, simply connect the white wire from SOURCE (which in this case is the white wire from outlet C) to the fixture as usual. The other wire on the fixture *must* be black, so connect the black wire of the cable that runs on to the first switch S-F-1. At the outlet, the white wire in the cable that runs on to the switch is re-identified as black and connected to the incoming black wire from source. The cable that runs on to the next outlet G must also be connected, black wire to black and white to white of the cable from source (from C). This completes the wiring of the outlet F. The switches still need to be connected.

Two different cables end in the box for S-F-1: a two-wire cable and a three-wire cable, making five wires altogether. Two of them are white; splice them together so there will then be a continuous white wire from F to S-F-1 to S-F-2, where you connect it to the common or marked terminal of that switch and re-identify as black wherever it is visible. Connect the black wire in the cable between F and S-F-1 to the common or marked terminal of the first switch S-F-1. That leaves two un-connected wires in the cable from S-F-1 and S-F-2: the red and the black. Connect them to the remaining terminals of each switch; it does not matter which color goes to which terminal on the switch. That finishes the wiring.

If you are installing four-way switches, follow the diagrams of Figs. 6–19 and 6–21. White wire may be re-identified where otherwise a wire of a different color would be required.

Outlet G is exactly the same as outlet B and should be wired in the same way.

Feed-through switch box   In all the outlets wired so far, you have run the cable first to the outlet box and fixture, then on to the switch. When you come to outlet H of Fig. 10–2, you will see that the cable runs first to the switch box S-H and then on to outlet H. This combination is even simpler to wire than the others because there

Outlet H

S-H

From G →

Fig. 10–7 Sometimes the cable from source does not run first to the outlet and then to the switch. Use this diagram when the cable runs first to the switch and then to the outlet.

is no problem with the colors of the wire, as you can see from Fig. 10–7 which shows this outlet completely wired.

Splicing for additional outlets   You *cannot* install additional outlets beyond H by simply connecting the black and white wires of the cable for the new outlet to the black and white wires in outlet H, because then the new outlet would be turned on and off by switch S-H. However, you can add an additional outlet by tapping in at switch S-H; splice the wires of the cable for the new outlet to the incoming cable from G, as shown in Fig. 10–8. Another way is to run three-wire cable from S-H to H; splice the two wires in the cable for the new outlet to black and white in H, as shown in Fig. 10–9.

A cable to any additional outlet can be spliced to any existing cable in any outlet box by splicing white to white, and black to black—as long as each wire can be traced all the way back to source without interruption by a switch.

Fig. 10–8 Tapping in and splicing to add an outlet beyond an existing outlet.

Fig. 10–9 Running three-wire cable and splicing to add an outlet beyond an existing outlet.

Three-way switches substituted for single-pole   It is a simple matter to substitute two 3-way switches for a single-pole switch in any wiring diagram. Study Fig. 10–10: the starting point is an outlet already wired with two wires ready for a switch. If a single-pole switch is to be used, connect it to the two ends of the wires as at A. If three-way switches are to be used, substitute the combination of B.

Fig. 10–10 It is a simple matter to substitute a pair of three-way switches for a single-pole switch in any wiring diagram. Simply substitute the right-hand combination for that at left.

Adding a switch to an unswitched outlet   If you want to add a switch to a diagram that shows an outlet permanently connected without a switch, cut the black wire. That gives you two ends of black wire to which the switch connects, or two ends to which you will splice the two-wire cable that runs to the switch.

Junction boxes   Sometimes it is necessary to make a T connection in cable when there is a long run and no convenient outlet from which to start the T branch. Use an outlet box; run the three (or more) ends of cable into it; splice all black wires together and all white wires together; cover with a blank cover as shown in Fig. 9–4A, and the job is finished. A junction box is shown in Fig. 10–11. *Remember: Junction boxes must always be located where permanently accessible.*

Fig. 10–11 A junction box contains only the splices of several lengths of cable.

Hot

Both halves hot

Top halves hot
Bottom halves contolled by switch

Grounded wire

Fig. 10–12 Two-circuit receptacles are well worth the extra cost.

TWO-CIRCUIT DUPLEX RECEPTACLES

It can be convenient to have a floor lamp or table lamp plugged into a receptacle controlled by a wall switch. These two-circuit duplex receptacles (also called split-wired receptacles) meet the NEC requirement for switch-controlled lighting in habitable rooms other than kitchens and bathrooms. Many brands of ordinary duplex receptacles are constructed so you can, during installation, change them to the two-circuit variety by breaking out a small brass portion between the two "hot" (brass colored) terminal screws. The wiring diagram is shown in Fig. 10–12.

THREE-WIRE CIRCUITS

In terms of material cost, installation time, and voltage drop, installing one 3-wire circuit offers many benefits over having two 2-wire circuits.

An ordinary 2-wire 120-volt circuit consists of two wires—the grounded wire and a hot wire as shown in Fig. 10–13. When two such 2-wire circuits are run to the same general area, there is a total of four wires as shown in Fig. 10–14 in which wires N_1 and N_2 are the grounded wires, and the wires A and B are hot wires *connected to opposite sides* of source, so that the voltage between A and B is 240 volts, but between A and N_1 and also between N_2 and B it is 120 volts.

Fig. 10–13 One 2-wire 120-volt circuit carrying 15 amps.

A   15 amps ⟶

15-amp load

⟵ 15 amps

N

Fig. 10–14 Two 2-wire circuits, fed from opposite legs of 120/240-volt service, each carrying 15 amps.

But note that the two grounded wires N_1 and N_2 are connected together at the neutral busbar in your service equipment. Since they are connected together at the starting point, they become in effect one wire. Why use two? There is no need for two. Use one wire and instead of two 120-volt 2-wire circuits you will have one 3-wire 120/240-volt circuit as shown in Fig. 10–15. Be sure that each receptacle is connected to the neutral wire to obtain 120 volts. If connected to the two hot wires the voltage would be 240 volts.

Fig. 10–15 One 3-wire circuit. Each half carries 15 amps but the neutral N carries no current.

Installing a 3-wire circuit not only saves time and material, but also reduces the voltage drop: if the loads on each half of the circuit (between A and N, or between B and N) are equal, the wire N carries no current at all. *Important:* In Fig. 10–14, each wire N_1 and N_2 is merely a *grounded* wire, but in Fig. 10–15, the single wire N will, if both loads are equal, carry no current at all, so it becomes a grounded *neutral* wire. If the loads are unequal, it will carry only the difference between the current in A and that in B, but it is still a neutral.

Where used   One 3-wire circuit is especially recommended in place of two 2-wire circuits for the small-appliance circuits discussed in Chapter 5. If the circuit is long, as for a yard light or as a feeder to a detached accessory building, voltage drop is a consideration (see pages 27–28). Other loads commonly served by 3-wire circuits are the dishwasher and waste disposer, the microwave oven and kitchen sink instant hot water heater, and the range and dryer. Even other general-use circuits often start out at the panelboard as 3-wire circuits to the first one or two outlets, where they split off to become two 2-wire circuits.

Connecting at the service   Great care must be taken to connect wires A and B (the hot wires) to opposite legs of the incoming service-entrance wires. Normally the wire N carries only a small number of amperes, or no current at all if the loads between A and N, and B and N, happen to be identical. But if you connect A and B to the same leg of the incoming service-wires, then wire N would be carrying a double load, possibly as much as 40 amps if No. 12 wire is used; its insulation would be damaged and fire might result. Remember the grounded wire is not protected by a fuse or breaker.

Three-wire circuits at receptacle and lighting outlets   When adding receptacles to a three-wire circuit, the neutral must be especially watched. A receptacle has two terminal screws on each side. When wiring two-wire circuits, it is the usual practice to end two different grounded wires at these two terminals, as is done especially when wiring with cable. But in wiring three-wire circuits, this must *not* be done because removing a receptacle, such as for replacement, would then result in a break in the neutral wire during the time there is no connection to the receptacle. This would in turn place all the receptacles connected to one leg (beyond the receptacle temporarily removed) in series with those connected to the other leg, all at 240 volts. Appliances would malfunction and be ruined or at least badly damaged, and lamps would burn out.

Maintaining a continuous ground in three-wire circuits   If using conduit, run a continuous wire with a loop as shown in Fig. 4–17. When using cable, there is no way of making a loop so you must make a pigtail splice as shown in Fig. 4–15. Connect the ends of the white wires of two lengths of cable to each other and to another short piece of white wire, the opposite end of which you then connect to the whitish terminal screw of the receptacle. It is good practice to pigtail both the grounded and hot wires at all receptacles where you would otherwise use the two terminals on each side for feeding through. In this way the removal of one receptacle does not interrupt service to other receptacles downstream, nor do the downstream receptacles depend on the terminal screws or push-in connections as a substitute for a good splice. Pigtail neutrals at lighting outlets also, so that replacing a fixture will not open the neutral to downstream outlets.

Installing split-wired receptacles on three-wire circuits   The split-wired (also called two-circuit) receptacle shown in Fig. 10–12 can also be installed on the three-wire circuit, wired with one half on phase A, the other on phase B (Fig. 10–15), and a common neutral. This is often done in the kitchen, where two small-appliance receptacle circuits are required. Be sure that the branch-circuit disconnecting means at the panelboard opens both hot wires with a single operation of the hand so that future maintenance at these outlets can be done safely.

DETACHED GARAGES AND ACCESSORY BUILDINGS

Lighting and receptacle requirements for garages and accessory buildings are discussed on page 14 as well as in this section on page 112. The feeder from the house to the detached building may be run underground or overhead.

Running underground or overhead wiring   If running the feeder underground you can use Type UF cable, which can be buried directly, or regular circuit wires with "W" in the name can be installed in a raceway such as rigid nonmetallic (PVC) conduit. The feeder must be buried at least 18 inches below grade. See further discussion in the chapter on farm wiring on page 175.

If you are going to use overhead wiring, there are several ways that the wires can be brought out of the house and into a separate garage or accessory building. Fig. 10–16 shows a very convenient entrance cap requiring only a single hole through the wall. The cap is used at both ends and is suitable for use with cable as shown. You can also use conduit by running it directly into the fitting, which accommodates ½-inch conduit.

Fig. 10–16 The easiest way of bringing wires into a garage or similar building is to use this special entrance cap.

Fig. 10–17 Another method of entering a building, using an ordinary entrance cap, conduit, and a box inside.

Fig. 10–17 shows another way of using an ordinary entrance cap (as used for service entrances), a short piece of conduit, and an outlet box inside the building.

Wiring for lighting and receptacles For a light in a garage or accessory building that can be controlled both there and at the house, three wires need to be run from the house and a three-way switch installed at each end, wired as shown in the basic diagram of Fig. 10–18 on the next page.

For receptacles that operate independently of the garage light so they are always live whether the light is on or off, four wires are required between house and garage. Follow the diagram of Fig. 10–19.

Fig. 10–18 Diagram for garage light fed by 3 wires from house with 3-way switch at each end.

Fig. 10–19 If a receptacle that is to be permanently on is added, 4 wires are needed.

Calculating feeder wire size Feeder wire size is based on the size of the load it must carry. To calculate the load of an accessory building that is a woodworking shop, machine shop, ceramic shop, greenhouse, etc., be sure to include the nameplate ratings of all the equipment—unless it is a one-person shop where only one tool will be used at a time. In that case, instead of adding all the tools together, just add in the tool with the highest ampere rating. Provide plenty of light, calculated at not less than 3 watts per square foot, and add 180 volt-amperes for each general-use receptacle. Don't forget outside lighting so you can safely get back to the house after dark. If the accessory building is some distance from the house, calculate the voltage drop, or use Table 4–3, so the feeder will not be undersized.

Grounding Where there is equipment in the garage or other detached building that requires grounding (metal boxes, grounding receptacles), either treat the electrical circuit or feeder to the garage as you would a service (as described in Chapter 8), installing a grounding electrode and a grounding electrode conductor, or run a separate equipment grounding conductor between the two buildings.

Where feeding one building from another, and where metal piping such as water pipe runs between the two buildings, the grounded conductor at the second building must be insulated from the enclosure and not connected to the water pipe or other grounding electrodes. In this case, a separate equipment grounding conductor is required to be run with the feeder. Any fault current returns to the service on this conductor.

CHAPTER 11
Cable—Nonmetallic-Sheathed and Armored

CABLE CONSISTS OF TWO OR MORE wires assembled together and then enclosed in a covering. The two most common types are nonmetallic-sheathed cable and armored cable. For descriptions of their characteristics, including coverings and size nomenclature, see pages 32–33. Approved locations and installation methods for the two types are described in this chapter.

NONMETALLIC-SHEATHED CABLE

Nonmetallic-sheathed cable is light in weight, simple to install without special tools, and costs less than other kinds of cable. It is often called "Romex," which is a trade name of one manufacturer. The National Electrical Code (NEC) permits it to be used only in one-family and two-family dwellings of any height, and in other buildings not exceeding three floors above grade. There are two kinds, which the NEC calls Type NM and Type NMC.

Where used Type NM Cable, shown in Fig. 4–5A and described briefly under that drawing, may be used only in *normally dry indoor locations.* Type NMC was developed for use in moist, damp or corrosive locations indoors or out, but not for exposure to the weather, and not to be buried in the ground. Type NMC may be used instead of Type NM anywhere. Its use is required in damp or corrosive locations such as barns and other farm buildings (further discussed in Chapter 17: *Farm Wiring*), damp basements, and so on. Install it as you would ordinary Type NM.

In some localities it may be difficult to locate Type NMC cable. In that case use Type UF, which is very similar, doesn't cost much more, and may be used wherever Type NMC may be used. In addition, it may be used in wet locations or buried directly in the ground. Type UF is discussed in more detail in Chapter 17: *Farm Wiring*, on pages 166 and 175.

Removing insulation Remove the outer jacket of the cable for 10 to 12 inches. You can use a knife, cutting a slit parallel to the wires and being careful not to damage the insulation of the individual wires. For Type NM, the cable ripper in Fig. 11–1 on the next page is very handy and faster than a knife.

Fig. 11–1 Using a cable ripper is the fastest way to remove insulation.

Fig. 11–2 Support cable at least once every 4 ½ feet. Use straps of this general type.

Bending   Bending cable sharply may damage the outer cover. The NEC says that all bends must be gradual so that, if continued in the form of a complete circle, the circle would be at least 10 times the diameter of the cable. Cable must always run in continuous lengths from box to box with no splices except inside boxes.

Fastening   Cable must be fastened to the surface over which it runs every 4½ feet, also within 12 inches of every box. Use insulated straps of the general type shown in Fig. 11–2, rather than the staples illustrated in Fig. 11–17 because it is too easy to damage the cable by driving staples too hard. Connectors and clamps should be snug but not over-tightened to the point of conductor damage. Sometimes such damage does not show up until later and is difficult to repair once the cable is concealed in the walls.

Whether exposed or concealed, cable may either follow the side of wood members or run through holes bored in the approximate center of studs, joists, or rafters. Where the edge of a bored hole or the nearest surface of a cable on the side is less than 1¼ inch from the face of the wood member, a ¹⁄₁₆-inch steel plate or a bushing must be installed to protect the cable against future penetration by nails, etc. Feed and pull the cable carefully through bored holes so as not to damage the outer jacket.

Protecting exposed wiring   If the finished wiring is exposed (for example in unfinished basements), see to it that the cable is protected against later mechanical injury. The easiest way is to run it along the side of a stud or joist. If run at angles to such timbers (unless the cable runs through bored holes), a running board must first be installed (see Fig. 11–3). The NEC is not specific as to the size of this running board but the so-called "1 by 2" is fine for the purpose. The cable must never be run across free space, and must follow the surface of the building except when mounted on running boards. In unfinished basements, it may be run through bored

Fig. 11–3 Exposed cable running across timbers must be protected by a running board to prevent damage.

Fig. 11–4 In attics, guard strips may be used. If run through bored holes, the cable needs no further protection.

holes through the center of joists (cables No. 8–3 or 6–2 or heavier may be mounted directly across the bottoms of joists without a running board). Where run parallel to framing members, run cable on sides, 1¼ inches or more from face. In accessible attics, cable may run at an angle to joists if protected by guard strips at least as high as the cable, as in Fig. 11–4. Only where necessary to provide protection against physical damage, Type NM cable can be drawn into any raceway, with a bushing installed on the raceway ends.

Use cable with grounding wire   Types NM and NMC are made both with and without a bare uninsulated (or green) grounding wire in addition to the two or three insulated circuit wires, but are seldom seen without the grounding wire. See Figs. 4–5A and 4–5B. In Chapter 10, the diagrams and explanations of wiring procedures are simplified by leaving out the grounding wire. But for most installations using nonmetallic-sheathed cable, you will use the kind with the bare uninsulated grounding wire. Connections for the bare grounding wire are explained below.

Where used   Cable with grounding wire is required to run to practically every box, though at one time it was required to run only to boxes containing receptacles. Since all receptacles in a dwelling must be of the three-wire grounding type, and all exposed metal parts of lighting fixtures and all metal boxes must be grounded, the only location where cable without a grounding conductor could be used would be to a nonmetallic fixture (such as the keyless porcelain or plastic socket in Fig. 9–4E) or to a switch in a nonmetallic box with a nonmetallic cover plate held on by nylon screws. It is more practical to use only cable with a separate bare (or green) grounding conductor to *every* box. The difference in cost does not justify keeping track of two kinds of cable on the job with the risk of using the wrong one.

Connecting the grounding wire   At the starting point in the service equipment, connect the bare grounding wire to the neutral strap, or separate grounding bus, in the cabinet. If the circuit originates in a separate panelboard that is *not part of the service equipment,* the bare grounding wire *must* be connected to a separate equipment grounding bus that is grounded to the cabinet and *not* to the neutral because the neutral is insulated from the cabinet at such locations. At all boxes, connect the insulated wires as if no grounding wire were involved. At each outlet or switch box, the bare wire must be properly connected. If two ends of bare wire enter the box, cut another piece of bare wire a few inches long or use a commercially available green insulated or bare pigtail assembled to a screw. Connect all three ends solidly together—you can use a solderless connector (Fig. 4–18). Ground the opposite end of the short wire to the box itself. You can use a clip (Fig. 11–5), or install an

Fig. 11–5 Clip for grounding to boxes.

Grounding wire

Wall of box

extra screw in one of the unused holes in the box. That screw may be used *only* for grounding the bare wire.

Receptacles in boxes   If a receptacle is to be installed in the box, run a bare wire (or insulated *green* wire) from the *green* terminal of the receptacle to the junction of the other bare wires. Your solderless connector will then join four ends of bare wire. See Fig. 11–6.

Fig. 11–6 How to install the bare grounding wire of nonmetallic-sheathed cable.

However, many UL listed self-grounding receptacles are now available with special mounting screws held captive by wire springs for connecting the receptacle to flush-mounted boxes, providing an effective bond between the receptacle yoke and the metal box. (When using nonmetallic boxes there is no need to connect the bare wire to the box, but all the bare wires must be connected to each other and to the green terminal of the receptacle.) If a receptacle is installed in a surface-mounted box such as a handy box in a basement, so that the mounting strap or yoke is in good solid contact with the box, the wire from the green terminal is not required. Regardless of circumstances, the bare wires in the cables must always be connected to each other and to the box. Be sure to fold all the bare wires well back into the box so they cannot later touch a live terminal on the receptacle.

Importance of a continuous ground   No matter what kind of boxes are used, the grounding wire must be installed so that it will be continuous from box to box all the way back to the service equipment. Removing a receptacle or other device from the box must not interrupt the continuity of the wire. This is in the interest of good grounding, which means a greater degree of safety. See the next paragraph, on "Testing the installation," for instructions on how to conduct a continuity test.

Testing the installation   Even experienced electricians who have confidence in their work test their installations. You should test your work before it is covered up and before receptacles, switches, and fixtures have been installed so that any corrections can be made while it is still accessible. Do your testing *before* the power supplier has energized your installation. There are two tests. The purpose of the first is to check for accidental grounds, and the purpose of the second is to make sure the conductors are continuous.

For both tests all you need is a continuity tester similar to that shown in Fig. 11–7. It is a "penlight" flashlight adapted for testing. Before each use, test the tester by touching the alligator clip on the flexible lead to the projecting prod at the lamp end. If the tester is working, the lamp should light.

Fig. 11–7 A continuity tester. Use only on de-energized circuits. It is also handy for testing fuses. *(M.W. Sperry Instruments Inc.)*

Testing for accidental grounds   Conduct this first test after the rough wiring is completed. In new construction that means work at the panelboard is completed and all permanent splices in the branch circuits are completed. Conduct this test *before* the power supplier has energized your installation. At the panelboard, be sure all the circuit breakers are off or the fuses removed. At each switch location, temporarily connect together the hot leg and the switch leg, either by using a solderless connector or just by hooking the bare wires together. Now go to each lighting or receptacle outlet and connect the alligator clip of the tester to the switch leg supplying the fixture (or the hot leg to the receptacle), and connect the prod on the tester to the equipment grounding conductor (there should be one at every outlet) as follows:

- If the box is nonmetallic—simply touch the tester prod to the bare or green grounding wire.

- If the box is metal—the grounding wire or cable armor should have been connected to the box, so simply touch the tester prod to the box.

In either case, the test light should *not* go on. This establishes that the hot wires throughout the job are "clean"—not accidentally grounded.

Testing for continuity   For this test, use a wire jumper (consisting of short wires that you cut for this purpose) to *temporarily* connect the two "hot" terminals (black, or black and red) and the neutral grounded (white) terminal together at the panelboard. Turn all of the circuit breakers on, or insert all the fuses. Now return to each outlet and use the continuity tester to test between the hot (black) wire and the grounded (white) wire at each outlet. The test lamp should light, showing there is continuity in both sides of the circuit back to the panelboard. When testing is completed, be sure to remove the jumper at the panelboard.

ARMORED CABLE

Armored cable (Type AC) is called "BX" by many people, though that is the trade-mark of one particular manufacturer. It is quite simple in construction (see Fig. 11-8). The wires are thermosetting or thermoplastic, each wrapped in a spiral layer of tough paper. The spiral armor is made of aluminum or galvanized steel and is strong but flexible. Between the paper and the armor there is a bare aluminum bonding strip or wire. This, together with the armor itself, serves the same purpose as the bare grounding wire in nonmetallic-sheathed cable. The armor itself is a very poor grounding conductor compared with copper because the spiral turns of the armor do not make good electrical contact with each other. The bonding strip is necessary to assure low resistance and a good ground.

Fig. 11-8 Armored cable consists of 2 or 3 wires protected by a layer of tough paper and flexible aluminum or galvanized steel armor. Note the bonding strip under the armor.

Another cable with interlocking metal tape armor, called metal-clad, Type MC, is similar in appearance to Type AC cable and can be used in the same locations and environments. Instead of supplementing the fault return path of the armor with an aluminum strip, Type MC cable includes a separate green insulated conductor for equipment grounding purposes.

Where used   Armored cable may be used only in permanently dry locations. Don't use it outdoors.

Cutting   The safest and quickest method of cutting armored cable is to use a patented cable cutter as shown in Fig. 11-9. It employs a circular saw blade to cut the armor, and has a stop to prevent cutting into the conductors. Make the cut about 10 inches from the end so you will have plenty of wire in the box for connections. If a cable cutter is not available, use a hacksaw, holding it at a right angle to the strip of

Fig. 11-9 A cable cutter offers the safest and quickest method of cutting armored cable. *(Seatek Co., Inc.)*

armor (Fig. 11–10). Be careful to saw only through the armor without touching the insulation of the wires or the bonding strip. This is not easy, so it helps to make some practice cuts on scrap cable. Try bending the cable sharply at the point where it is to be cut. It is difficult to saw completely through a single turn of armor, but usually if the center part of the strip is sawed completely through and the edges partly through, a sharp bend will break the armor and a sharp twist will remove the short end (see Fig. 11–11).

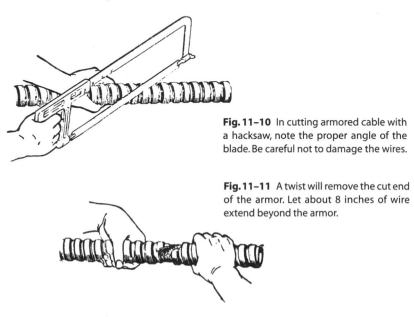

Fig. 11–10 In cutting armored cable with a hacksaw, note the proper angle of the blade. Be careful not to damage the wires.

Fig. 11–11 A twist will remove the cut end of the armor. Let about 8 inches of wire extend beyond the armor.

Anti-short bushings and connectors   If you cut a length of cable and examine the cut end of the armor, you will find sharp, jagged edges pointing inward toward the wire. These teeth tend to puncture the insulation of the wire and cause short circuits and grounds. The NEC requires a bushing of tough fiber or plastic to be inserted at the end of the cable between the steel armor and the wires. They are called "anti-short bushings" and are supplied with the cable (see Fig. 11–12).

In some cables an overall paper wrapping may make it difficult to insert the anti-short bushing. To make room for it, unwind the paper a few turns under the armor,

Fig. 11–12 Always use a fiber or plastic anti-short bushing at the end of the cable between wires and the armor. It provides protection against the sharp edges of the armor that might puncture a wire and cause a ground or a short circuit.

then give it a sharp yank and it will tear off inside the armor (see Figs. 11–13 and 11–14). Then insert the anti-short bushing as shown in Fig. 11–15. Fig. 11–16 shows a cross-section of a piece of cable with the anti-short bushing properly inserted.

Leave an inch or two of the bonding strip projecting beyond the end of the armor. Insert the anti-short bushing as described above, then fold the bonding strip over the bushing and back along the outside of the armor. The bonding strip will help hold the bushing in place until the connector is installed.

Then install a connector as shown in Figs. 9–12 and 9–13 (unless you are using boxes with cable clamps, which make the use of cable connectors unnecessary). The connectors used with armored cable are similar to those used with nonmetallic-sheathed cable except that at the end which goes into the box there are "peepholes" through which the color of the anti-short bushing can be seen. This allows the inspector to see that the bushings have been installed. If you do not use the bushings, your job will not be passed by the inspector.

Fig. 11–13 In removing paper, first unwind a few turns inside the armor.

Fig. 11–14 A sharp yank tears off the paper inside the armor, and makes room for the anti-short bushing.

Fig. 11–15 Insert the anti-short bushing between wires and armor.

Fig. 11–16 A properly installed anti-short bushing.

For Type MC cable, the anti-short bushing isn't required at the cut ends of the armor, but some manufacturers furnish them with each coil of cable, and it is a good idea to use them. Fittings must be identified as suitable for use with Type MC cable.

Installing armored cable After you have prepared the ends of the armored cable with bushings and connectors, install the cable as described for nonmetallic-sheathed cable on page 114. Like nonmetallic-sheathed cable, armored cable must be supported every 4½ feet and also within 12 inches of each box. It may be supported using straps (Fig. 11–2) similar to those used with nonmetallic-sheathed

Fig. 11–17 Staple for supporting armored cable.

cable, but staples are more frequently used; see Fig. 11–17. The NEC requires the staples to be rust-proof. Drive the staples home with a hammer, but not hard enough to damage the cable.

For armored cable, the problems concerning correct colors of wire are exactly the same as with nonmetallic-sheathed cable. As with nonmetallic cable, splices are not permitted. The cable must be in one piece from box to box. If a splice is necessary, make it in a junction box as shown in Fig. 10–11.

Anchoring for grounded connections Be especially careful in anchoring the connectors to outlet and switch boxes. The connector must be tightly clamped to the armor of the cable. Be sure the aluminum bonding strip of the cable is bent back over the outside of the armor and solidly clamped by the connector. In other words, don't let the strip lie in a part of the connector that doesn't solidly clamp it against the armor. After inserting the connector into the knockout of the box, drive the locknut down tightly enough to bite down into the metal of the box. This makes a good electrical connection so that, in case of accidental grounds, current can flow through the armor from box to box. The white wire in the cable is grounded. The grounding strip is also grounded. If the black wire at some point where the insulation is removed accidentally touches the armor or the box, the result is the same as touching the white wire. It is a ground fault and causes the fuse protecting that circuit to blow or the circuit breaker to open—a signal that something is wrong.

Grounding terminals of receptacles in boxes The green grounding terminal of the receptacle must be *effectively* grounded. But with armored cable there is no need for the extra grounding wire that is part of nonmetallic-sheathed cable. Instead, the armor of the cable and the bare aluminum grounding strip under the armor serve the same purpose.

If the mounting yoke of the receptacle is in firm and solid metal-to-metal contact with the box, as in surface wiring, no further action is necessary. In flush work the plaster ears on the yoke of the receptacle usually prevent it from resting directly on the box. As a result the only metallic contact between the box and the receptacle is through the small mounting screws, and that is not good enough. Unless you use the specially approved self-grounding receptacles designed for use without the grounding wire in metal boxes (see page 116), you must install a short length of wire from the green terminal of the receptacle to the box, using either of the methods discussed for nonmetallic-sheathed cable on page115.

Testing the installation Perform the same tests described earlier in this chapter on pages 116–117. In the test for accidental grounds, use the second test (starts with "If the box is metal...") which calls for touching the tester prod to the metal box, because only metal boxes are used with armored cable.

CHAPTER 12

Conduit

CONDUIT IS A PROTECTIVE wiring enclosure (raceway) through which conductors are pulled after the conduit is installed. The wires and cable may also be removed from conduit if necessary. The most commonly used types of conduit are rigid metal conduit, rigid nonmetallic conduit, intermediate metal conduit (IMC), electrical metallic tubing (EMT), and flexible metal conduit (also referred to as "flex" or by the trade name "Greenfield").

This chapter explains general installation procedures, followed by details specific to each of the five conduit types. Space does not permit description of other types such as liquidtight flexible metal conduit, which is covered in Article 351 of the National Electrical Code (NEC), or electrical nonmetallic tubing (NEC Article 331).

CHOOSING AND INSTALLING CONDUIT

For size of conduit, there are specific NEC requirements for each type, while the general information given below about box depth, bending, and anchoring applies to several types of conduit.

Calculating correct conduit size   The size of conduit you will need to use is determined by the conduit type and the dimension and number of wires. Whether electrical metallic tubing, flexible conduit, intermediate metallic conduit, or rigid conduit is used, Tables C1, C3, C4, and C8 in Appendix C of the NEC restrict the number of wires that may be used inside each size and type of conduit. These fill tables occupy 22 pages in the NEC, so they cannot be reproduced here.

To calculate the minimum size raceway for combinations of different sizes of wires or types of insulation in the same raceway, refer to Chapter 9 in the NEC. Use Table 1 to find the permitted percentage of fill in a raceway, Table 4 to find dimensions of conduit, and Table 5 to find dimensions of insulated wires. Again, these NEC tables are too lengthy to reproduce here.

Box depth with all conduit types   When installing any kind of conduit, be sure to use boxes deep enough so that the knockouts on the sides or ends of the box are close enough to the back to allow room for the conduit behind the thickness of the wall or ceiling finish material.

Bends in conduit   No run (as the conduit from box to box is called) may have more than the equivalent of four quarter-bends (a total of 360°) in it. Until you gain experience you may find it best to bend the conduit first, then cut to the required length, because the finished piece after bending may turn out to be a fraction of an inch too long or too short.

Anchoring with pipe straps   Support the conduit using pipe straps. It must be supported within 3 feet of every box, with additional supports not more than 10 feet apart for electrical metallic tubing and for ½-inch and ¾-inch rigid metal conduit and intermediate metal conduit. You can find support spacing requirements for larger rigid and intermediate metal conduits in Table 346-12 of the NEC.

Wires in conduit   All wires must be continuous, with no splices permitted inside the pipe. The conduit is first installed, and the wire is pulled into it later. See page 126 for instructions on pulling wire into conduit.

FOUR TYPES OF CONDUIT

Location requirements and other details of four types of conduit are described here. A fifth type, flexible metal conduit, appears at the end of the chapter.

Rigid metal conduit   Generally steel, though it may be made of aluminum or other metals, rigid metal conduit (Fig. 12–1) differs from water pipe in that it is carefully inspected to make sure the inside is entirely smooth to prevent damage to the wires as they are pulled into the pipe. It has a rust-resistant finish inside and outside. Each length bears a UL label.

Where used   Rigid metal conduit may be used either indoors or outdoors, but it may require supplementary protection in corrosive locations such as in cinder fill.

Sizes   It has approximately the same dimensions as equivalent sizes of standard weight water pipe. Standard sizes of rigid metal conduit are ½-inch, ¾-inch, 1-inch, 1¼-inch, 1½-inch,

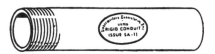

Fig. 12–1 Rigid metal conduit looks like water pipe but differs in several important aspects. It comes only in 10-foot lengths and each length bears a UL label.

Fig. 12–2 A conduit bender is needed to make good bends in conduit. The one shown here is for EMT. Instructions usually come with the bender.

2-inch, and 2½-inch, plus larger sizes used mostly in commercial work. The actual inside diameter is a little larger than the trade sizes indicated above.

Bending   If conduit is bent sharply, it will collapse. Bends must be gentle and gradual so the internal diameter will not be reduced at the bend. The NEC requires bends which, if continued into the form of a complete circle, would be at least 8 inches in diameter for the ½-inch size; 10 inches for the ¾-inch; 12 inches for the 1-inch; 16 inches for the 1¼-inch; 20 inches for the 1½-inch; 24 inches for the 2-inch; 30 inches for the 2½-inch. To do a good job, use a bender similar to the one shown in Fig. 12–2. For larger sizes, use factory-bent elbows.

Cutting and threading   The preferred way to cut rigid metal conduit is to use a hacksaw with a blade having 18 teeth to the inch. Cutting will leave a sharp edge or at least burrs at the cut, which might damage wires as they are pulled into the conduit. Use a pipe reamer to remove these dangerous projections. Thread the conduit using dies similar to those used with water pipe, with a taper of ¾ inch to the foot.

Anchoring with locknut and bushing   Conduit is anchored to a box by means of a locknut and bushing (see Fig. 12–3). The locknut is used on the outside of the box, the bushing on the inside (see Fig. 12–4). The locknut is not flat, but has teeth on one side; the side with the teeth faces the box. The bushing has a rounded surface on the inside diameter, over which the wires slide while being pulled into the pipe.

Fig. 12–3 Conduit is fastened to boxes by means of locknuts and bushings, both shown at left.

Fig. 12–4 This shows how a locknut and bushing are used. Be sure the locknut is driven solidly home.

To install properly, first screw the locknut on the pipe as far as it will go, but don't do the final tightening yet. Next slide the pipe through the knockout in the box. Install the bushing on the inside of the box, screwing it on tight as far as it will go. Finally, tighten up the locknut on the outside of the box making sure the teeth dig down into the metal of the box.

Grounding   It is important to tighten the locknut as just described in order to provide a good continuous ground. Many electricians prefer—and some local codes require—that two locknuts be used, one on each side of the box wall, plus a bushing. The NEC requires double locknuts and no concentric or eccentric knockouts where circuits are over 250 volts to ground.

Rigid nonmetallic (PVC) conduit   There are many types of nonmetallic conduit. Impregnated fiber, fiberglass epoxy, and polyethylene are used primarily underground. Only polyvinylchloride (PVC) is discussed here. It is light in weight, easy to install, and has excellent moisture and corrosion resistance. When calculating fill for PVC, be sure to include the equipment grounding conductor, required in essentially every run. See NEC Article 347 for further details.

Where used   Except for some hazardous locations and for support of fixtures, PVC can be used wherever rigid metal conduit is permitted, and is preferred for some corrosive locations.

Sizes   PVC conduit in "Schedule 40" has the same dimensions as rigid metal conduit, and in the heavier-wall "Schedule 80" it has the same outside dimensions, but reduced inside dimensions.

Bending   Couplings and fittings are solvent-cement "welded" to the outside of the conduit. Factory-made bends and elbows are available, or the conduit can be field bent by first applying heat to soften the material and then using a jig to form the bend, followed by applying a wet cloth to cool it. There are several commercially produced bending heaters available using liquid, air, or convection. In the larger sizes, steps must be taken to maintain the round cross section at bends so it does not flatten.

Cutting   Any fine-toothed saw can be used to cut the material. The cut ends should be reamed inside and out.

Intermediate metal conduit (IMC)   Made only of steel, intermediate metal conduit has slightly smaller wall thickness and larger inside diameter than rigid metal conduit.

Where used   In general, IMC is recognized for the same uses as rigid metal conduit.

Cutting and bending   Cutting, threading, and bending are done the same as for rigid metal conduit, except that benders providing side wall support, similar to the EMT bender shown in Fig. 12–2, must be used because the thinner walls must be supported to prevent collapse during bending. Threaded ends will appear a little different, as the tops of the first few threads will be flat, but properly cut threads will mate with standard conduit fittings.

Electrical metallic tubing (EMT)   This is also called thin-wall conduit. It is shown in Fig. 12–5. Each length bears a UL label.

Where used   EMT may be used either indoors or outdoors.

Sizes   In trade sizes ½-inch through 1½-inch, it has the same *inside* diameter as rigid metal conduit. The *outside* diameter for these sizes is much smaller than for rigid conduit because the wall is much thinner; in fact the wall is so thin that it cannot be threaded. In sizes 2-inch and larger, EMT has the same outside diameter as rigid metal conduit.

Cutting and bending   A hacksaw with 32 teeth to the inch is the most convenient tool for cutting EMT. The tubing must be reamed after cutting. Bend it as you would rigid metal conduit.

Fig. 12–5 EMT cannot be threaded. It is much lighter than rigid conduit. It comes only in 10-foot lengths.

Anchoring with couplings and connectors   Lengths are coupled together and connected to boxes with special pressure fittings—coupling and connector of one type are shown in Fig. 12–6. Be sure the end of the tubing goes all the way into the connector, against the shoulder stop. Tightening the nut securely clamps the tubing into the fitting.

Fig. 12–6 EMT is anchored to boxes using the connector shown at left. A coupling is shown at right.

INSTALLING WIRES IN CONDUIT

Assume that you have installed the conduit for outlets A, B, C, and D of Fig. 10–2 in Chapter 10. The installation will look as shown in Fig. 12–7. You are ready to pull the wires into place. For a short run with just two small wires, they can probably be simply pushed in at one end and through to the next outlet. If the runs are longer or have bends, fish tape must be used.

Fig. 12–7 Outlets A, B, C, and D of Figs. 10–1 and 10–2 installed with runs of conduit and ready for wires to be pulled in.

Using fish tape   Fish tape is a highly tempered steel tape about ⅛ inch wide and ¹⁄₁₆ inch thick. As purchased, it will have a loop formed at one end. The tape is flexible enough to go around corners, but stiff enough not to buckle when pushed into conduit. Push the tape into the conduit until the loop appears where the wires are to enter the conduit. Push the bare wires through the loop of the tape, then twist them back upon themselves as shown in Fig. 12–8. Insulating tape may be used as shown to help hold the wires. If the loop on the end of the tape breaks, you can't form a new one by just bending, because the highly tempered tape would break. Heat the end of the tape with a blowtorch, then let it cool; this will soften the end so a new loop can be formed.

Fig. 12–8 Fish tape makes it easy to pull wires into conduit.

Tape optional

Wiring at boxes In pulling the wires into outlets of Fig. 12-7, pull in short pieces from A to B, separate pieces from B to S-B, from B to C, and from C to D. But if the wires from A to B to C, for example, do not require connections at B, you would feed a length in at A, through B, and on to C, leaving a loop in the box at B to facilitate future changes. At each box where a connection is to be made, leave 10 to 12 inches of wire beyond the box. Outlet S-B will contain a switch; use only black wire because NEC Section 200-7 permits a re-identified white wire (explained on page 103) *only in cable* to be run as the supply to a switch.

Grounding receptacles Ground the green terminal to the box unless you are using the self-grounding receptacles mentioned on page 116 that makes this unnecessary.

Testing the installation Test as for armored cable as described at the end of the previous chapter.

FLEXIBLE METAL CONDUIT

Flexible metal conduit, shown in Fig. 12–9, is available in both steel and lightweight aluminum. It is commonly referred to as "flex" or by the trade name "Greenfield." Flexible metal conduit is similar to the armor of armored cable but the wires are pulled in after the flex is installed.

Fig. 12–9 Flexible conduit is installed in the same way as armored cable, but wires are pulled into place later.

Where used Use only indoors in permanently dry locations (unless using wire approved for wet locations: one having a "W" in the type designation, such as Type TW). Flexible metal conduit is not commonly used for a complete wiring system except in a few communities, but it is widely used where some flexibility or movement is required, such as for motors.

Sizes   The smallest for general use is trade size ½-inch, but it has about a ¾-inch outside diameter.

Anchoring with connectors   Install flexible metal conduit as you would armored cable. Use connectors of proper size, and then pull the wires into place.

Grounding   You must also install a grounding wire, which may be bare, or insulated if green. But in lengths up to 6 feet, where flexibility is not required and vibration is not present, and where the conductors are protected at 20 amps or less, the flex may serve as the equipment grounding means without an additional grounding wire. The size of the grounding wire depends on the rating of the fuse or breaker protecting the hot wires. If 15-amp, use No. 14; if 20-amp, use No. 12; if 30-amp to 60-amp, use No. 10; if 100-amp, use No. 8; if 200-amp, use No. 6. This wire is connected just as the bare grounding wire is connected in nonmetallic-sheathed cable. Ground the green terminals of receptacles as with nonmetallic-sheathed cable.

Testing the installation   Test as for nonmetallic-sheathed cable and armored cable as described on pages 116–117 in the previous chapter. In the test for accidental grounds, use the second test (starts with "If the box is metal . . .") which calls for touching the tester prod to the metal box.

CHAPTER 13
Modernizing Old Wiring

THE WIRING THAT WAS INSTALLED in a house many years earlier, or even as recently as a decade ago, may not be adequate for the job it is called upon to do today. A complete rewiring job is in order—or is it? Do not jump to the conclusion that every outlet must be torn out and every receptacle replaced. Many times a less expensive job will serve the purpose.

Is the wiring inadequate because you are using too many lights? too many floor lamps? too many radios and TVs? That is seldom the case. The wiring usually is inadequate because you have added many electrical appliances that were not considered or perhaps were not even on the market at the time of the original wiring job. The installation does not provide enough circuits to operate a wide assortment of small kitchen appliances, plus range, water heater, clothes dryer, room air conditioners and other heavy appliances. Some of these operate on 240-volt circuits, which may not be available; others operate at 120 volts but when plugged into existing circuits they overload those circuits. In addition, the service entrance equipment may be just too small for the load.

To analyze the problem of your particular house, ask yourself this: If you disconnected all the *appliances,* would you have all the *lighting* circuits you need? The answer is probably yes, which means that your rewiring job is simplified. You will still have to rewire the house, but probably not as completely as at first appeared necessary. Proceed as if you were starting with a house that had never been wired, but leave the existing lighting circuits intact. (These lighting circuits, of course, will include many receptacles used for small loads like a vacuum cleaner, radio, and TV, but not the receptacles for kitchen or laundry appliances.)

This chapter discusses the overall plan for modernizing an existing installation, including upgrading to a larger service and adding new circuits. It also offers problem-solving approaches to the specific challenges you are likely to encounter. Everything in previous chapters describes the wiring of buildings while they are being built, which is called "new work." This chapter describes "old work," which is the wiring of buildings after they have been completed.

PROBLEMS OF "OLD WORK"

There is little difference between old and new work, except that in old work there are a great many problems of carpentry. The problem is to cut an opening where a fixture is to be installed, and another where a switch is to be installed, and then to get the cable inside the wall from one opening to the other with the least amount of work and without tearing up the walls or ceilings more than necessary.

One house to be wired may be five years old, another a hundred years old. Different builders use different methods of carpentry. Every job will be unique. No book can possibly describe all the methods used and all the problems you will meet. Watch buildings while they are being built to get an idea of construction at various points. In old work, good common sense is of more value than many pages of instruction.

In general, old work requires more material because it is often wise to use ten extra feet of cable to avoid cutting extra openings in the walls or to avoid cutting timbers. Many problems can be solved without cutting any openings except the ones to be used for outlet boxes and switch boxes. Others require temporary openings in the wall which must later be repaired. Techniques are given in this chapter for running cable behind walls and ceilings, and for installing boxes, switches, receptacles, and outlets in both lath-and-plaster and drywall construction.

Caution: Even in an abandoned, vacant, or isolated building, always test for voltage before starting any work. It is possible, with other buildings on the property, or with a two-meter service for yours and another occupancy, that there might be feedback which would be surprising at the least, or hazardous at worst.

WIRING METHODS

The wiring method you will use in old work will be either nonmetallic-sheathed cable with grounding wire, or armored cable—whichever is the custom in your locality. The rigid conduit or the EMT methods cannot be used unless the building is being practically rebuilt. In a few areas, flexible metallic conduit is used. In this case be sure to install the grounding wire along with the circuit wires as discussed on page 128.

Knob-and-tube wiring is permitted by the National Electrical Code (NEC) Article 324 only for extensions to existing work. The materials (knobs, tubes, and loom) are difficult to find and the building wires currently manufactured do not fit the knobs. If you need to extend a knob-and-tube circuit, or pick up a portion with a new house run, extend the wires encased in loom (a flexible woven fibrous nonmetallic tubing) continuous from the last knob into a new accessible junction box, and change to the cable wiring method you are using.

INSTALLING NEW SERVICE AND CIRCUITS

Install a new service of at least 100-amp capacity, with a new outdoor meter, beginning with the insulators on the outside of your house and ending with new service equipment inside the house. Let the equipment contain, as a minimum, branch circuit breakers to protect all the new circuits that you are going to install, plus a few spares. Be sure to consult your power supplier about how you plan to proceed. Review Chapter 8 for calculating service size.

When you have finished installing your new service and circuits, you will have no power on your new circuits but will still have power on the old circuits. *Call your power supplier to disconnect all power on the outside of your house.* Get along for a day or two without electric power while you reconnect the old circuits into your new equipment.

When all the work has been completed, have your power supplier install the meter and connect the power to your new service. Then enjoy the advantages and pleasures of a newly rewired home.

Installing new circuits   Using the old-work techniques described later in this chapter, install two small-appliance circuits, or better, one 3-wire circuit. Install the laundry circuit and the bathroom receptacle circuit. Install individual circuits for heavy appliances such as range, water heater, clothes dryer, and furnace motor. You will be connecting each of these circuits to the breakers in the new equipment.

Installing the service equipment   Whether you retain or replace the old service equipment panelboard, you must disconnect the incoming service wires from the present equipment. Follow Option A below if you will continue to use your old panelboard. Follow Option B if you plan to discard your old panelboard entirely. With either option, you must disconnect the ground wire from the old equipment and remove it. You must install a new ground wire from the new equipment to the ground, and an equipment grounding conductor between the new and old equipment enclosures. In the discussion that follows, circuit breakers are mentioned but fused equipment may be used.

Option A—Using the old panelboard   Your new equipment will contain a main breaker, breakers for all the new branch circuits you have installed, plus a few spares for future circuits, and one 2-pole 30-amp or larger breaker from which you will run wires to the old equipment. Let's assume your present installation is 3-wire 120/240-volt.

Disconnect the incoming service wires from your old equipment. If they are in conduit, remove the conduit. If they are in cable, remove it so it doesn't enter the cabinet at all. Then run three-wire cable from the old equipment to the new. The two hot wires of the cable run from the 30-amp or larger breaker mentioned in the previous paragraph to the terminals in the old equipment to which the old service wires were connected. The white wire in the cable runs from the grounded busbar in your new equipment to the old equipment. The white wires in your old equipment will present a problem. If the old equipment was installed comparatively recently, the grounded busbar in it may be bonded to the cabinet either (1) by a special bonding screw: remove it and the busbar will then be insulated from the cabinet; or (2) by a flexible metal strap bonded to the cabinet with the other end connected to the grounding busbar: disconnect it from the grounding busbar, which will then be insulated from the cabinet (it will be best to cut off the bonding strap completely). Then connect the white wire from the grounded busbar in the new equipment to the now-insulated busbar in the old equipment.

If your old equipment was installed many years ago before the schemes described in (1) and (2) above were in use, the grounded busbar was probably bonded directly to the cabinet with no way of insulating it. Remove it if possible, but in any case remove the white wires from it and install them in a new, insulated, grounded busbar which you must purchase and install. Make sure it has enough terminals of the right size to accommodate all the white wires.

If your old equipment is 2-wire 120-volt, the procedure is the same as outlined in the preceding paragraphs except that your new equipment, instead of containing one 2-pole breaker to protect the old branch circuits, will contain a single-pole breaker for that purpose. Run a 2-wire cable, instead of a 3-wire, from the new equipment to the old.

This completes the wiring. Your old equipment is now connected to the new equipment. If your old equipment contained *main* breakers or fuses, leave them as they are; they are not required but will do no harm. But if there are fuse clips that seem deteriorated or terminals that appear to be in poor condition, remove them.

If the original equipment served only lighting circuits (including receptacles), the wires from the old to the new equipment may be No. 10 protected by a 30-amp breaker. But if the old equipment served and continues to serve 240-volt loads as well, install No. 6 wires protected by a 60-amp breaker. See Fig. 13–1.

Fig. 13–1 Connections using Option A for retaining the old panelboard.

Option B—Replacing the old panelboard Remove the original service wires. Disconnect all branch circuit wires, but don't cut off any of the wires. If the trim or cover of your present equipment has a hinged door in it, or if any openings in it can be closed off, remove and discard the interior (breakers and bus, or fuseholders, terminals, neutral bar, etc.), install knockout seals in any unused openings, and use the enclosure as a junction box. If this is impractical, proceed as follows: if the wiring used armored or nonmetallic-sheathed cable, remove the locknut inside the cabinet, pull the cable connector out of its knockout, and temporarily screw the locknut on its connector. If the wiring was in conduit, remove the bushing inside the cabinet, pull the conduit out of the knockout, and place the bushing on the end of the conduit. After doing this on each branch circuit, remove the old equipment completely.

The new equipment will contain a main breaker, plus other breakers to protect each of the branch circuits, old and new, plus a few spares. But the wires of the old circuits will not reach the new equipment. Where the old equipment used to be located, install a junction box (an empty steel cabinet of convenient size, 8 × 8 inches or larger as needed, with steel cover). Run the wires of the old circuits into this junction box, using the original cables with their connectors, or the original conduits.

Fig. 13-2 Connections using Option B for replacing the old panelboard.

Then run separate cables from the new equipment to the junction box, one for each circuit and of the same size as the original wires. Connect the new wires to the old, black to black, white to white, using "wire nuts" or other solderless connectors. See Fig. 13-2. The junction box will not contain a grounding busbar, and the white wires in the box must be carefully insulated from each other and totally insulated from the box, but the box itself must be grounded.

INSTALLING BOXES AND RUNNING CABLE

The NEC requires that boxes enclosing flush devices must be at least $^{15}/_{16}$ inch deep. Lighting outlet boxes may be as little as ½ inch deep as shown in Fig. 13-3, but deeper boxes should be used wherever possible. Cable is simply pulled into the walls and anchored to the outlet and switch boxes. Each piece must be a continuous length from box to box. In all wiring the NEC limits any gaps between boxes

Fig. 13–3 The NEC permits boxes as shallow as ½ inch, which is the type shown here.

and the surrounding wall material to ⅛ inch or less. If it is not practical to cut openings close to the exact size of the box, use patching plaster to fill gaps after the wiring is finished.

Selecting box location   In locating outlets, consider that all wires must be fished through walls and ceilings. Sometimes by moving an outlet or switch a foot or so, a difficult job of boring through joists or other timbers can be avoided. In old work, the switch and outlet boxes are supported directly or indirectly by either the lath under the plaster or by the drywall, so choose the locations for the openings carefully. A location fairly close to joists and studs is best because there the wall materials are better supported.

Mounting switch boxes in lath-and-plaster construction   In sawing openings for switch boxes in walls and ceilings of lath-and-plaster construction, remember that the length of a switch box is approximately the same as two widths of lath plus the space between the laths. If you remove two complete widths of lath, the mounting brackets on the switch boxes will barely reach the edges of the next two laths, and the laths will split when you drive the screws (1-inch No. 5 screws are commonly used). Remove one width of lath completely, and part of another on each end of the opening. Fig. 13–4 shows the wrong and the right way.

Fig. 13–4 In cutting an opening for a switch box, cut away one whole lath and part of another lath on each side of the one completely cut. This provides a rigid mounting for the box.

Make a mark on the wall approximately where the switch or receptacle is to be located. Bore a small hole through the mark. Insert a stiff wire and probe to make sure there is no obstruction and that there is sufficient space all around. Enlarge the hole to locate the center of one lath—this will locate the vertical midpoint of your opening. Then mark the area of your opening, about 2 inches by 3¾ inches.

Bore ½-inch holes at opposite corners, and at the center of top and bottom. See Fig. 13–5. The centers of the holes must be on the lines of the outline so that not more than half of each hole will be outside of the rectangle. Unless you watch this carefully, parts of the holes may later not be covered by your switch or receptacle plate.

The holes at the corners are for inserting a hacksaw blade or fine-toothed keyhole saw for cutting the opening. The holes at top and bottom provide clearance for the screws used to mount switches or receptacles. If using a hacksaw blade, wrap one end with tape for a handle. Insert the blade so the sawing is done as you pull the blade out of the wall. Support the plaster with a piece of wood as you pull the blade toward you. If you saw as you push the blade into the wall, you will probably loosen the laths from the plaster, leading to a very flimsy mounting of the box.

Mounting switch boxes in drywall construction   Drywall is not sturdy enough to accept screws as is lath-and-plaster construction. Mount the box near a stud (use a stud finder) for better support. Use the template in Fig. 13–5 to cut the opening. Adjust the brackets on the ends of the box so that when installed the box will be flush with the wall surface.

Fig. 13–5 Dimensions of hole for installing a switch box. To save time, make a template. Photocopy this page, paste to a piece of stiff cardboard, and cut to size.

3 1/4 in.

2 in.

Connecting the cable   In both types of construction, boxes with beveled corners and cable clamps are preferable but not essential. Bring the cable into the box and tighten the clamps, letting about 10 inches of cable extend out of the box. If using boxes without clamps, install the connector on the cable, let the connector project into the box through a knockout near the back of the box, and install the locknut after the box is installed. Be sure the box is deep enough and the knockout far enough back for the cable connector to clear the inner surface of the wall.

Anchoring switch boxes   The following three methods can be used in both lath-and-plaster and drywall construction:

▪ Use a box that has special clamps on its outside walls (see Fig. 13–6). After installing the cable, push the box into the opening, then tighten the screws on the external clamps. This makes the clamps collapse, anchoring the box in the wall.

▪ Use an ordinary box plus the U-shaped clamp of Fig. 13–7. Install the U-clamp with the screw holding it in place unscrewed about as far as it will go. When you slip the box into its opening, the ends of the clamp will expand outward. Tighten the screw holding the clamp to anchor the box ears firmly against the wall.

▪ Use an ordinary box plus a pair of special straps shown in Fig. 13–8. Insert one strap on each side of the wall opening and push the box into the opening, taking care not to lose one of the straps inside the wall. Then bend the short ends of each strap down into the inside of the box. Be sure they are bent sharply over the edge of the box and lie *tightly* against the inside walls of the box so they cannot touch the terminals of a switch or receptacle installed in the box, which would lead to grounds or short circuits.

Fig. 13–7 Another way of supporting switch boxes.

Fig. 13–6 This type of box saves time in old-work installations.

Ear of box

Fold inside of box

Box

Wall board

Fig. 13–8 A pair of these straps supports a switch box.

Mounting outlet boxes in ceilings   If there is open space above the ceiling on which the box is to be installed, and if there is no floor above (or there is a floor in which a board can easily be lifted), proceed as in new work using a hanger and the usual 1½-inch-deep box. See Figs. 9–8, 9–9, 9–10. The only difference is that you will be working from above.

If all the work must be done from below, the method depends on the ceiling construction, the location of the outlet, and the weight of the fixture. Boxes supporting fixtures weighing more than a few pounds must be fastened to the building structure. One method is shown in Fig. 13–9. For lightweight fixtures the ceiling itself can support the box. Use one of these methods: (1) ½-inch-deep box, surface mounted, supported by a ceiling joist using screws; (2) ½-inch-deep box, surface mounted, supported by fixture stud on bar hanger (Fig. 9–8) which has been poked up through a hole (which the box covers) and laid across the wood laths (Fig. 13–10); (3) ½-inch-deep box, surface mounted, supported by toggle bolts through ceiling, either lath and plaster or drywall (Fig. 13–11); (4) 1½-inch-deep box with ears, flush mounted, supported by a U-clamp (Fig. 13–12). When using a surface-mounted ½-inch-deep box, be sure to select a fixture having a canopy which will cover the box, as shown in Fig. 13–9. In all these cases, fish the cable to the outlet location and fasten it to the box before securing the box.

Remember that if the ceiling is combustible (if it can burn), you must cover the space between the edge of the box and the edge of the fixture with a noncombustible material.

Fig. 13–9 Surface-mounted ½-inch-deep box fastened to ceiling joist.

Fig. 13–10 Surface-mounted ½-inch-deep box supported from bar hanger by long fixture stud.

Fig. 13–11 Surface-mounted ½-inch-deep box supported from drywall toggle bolts.

Fig. 13–12 Recessed box supported from drywall board by U-clamp and box ears.

Lifting floor boards   Often a board in the upstairs floor must be lifted to get at the ceiling space. This is no problem if the flooring is rough, as in ordinary attics. But if the lumber is tongued and grooved, care must be taken to avoid marring the floor. The first step is to cut the tongue off the boards. A putty knife cut off short, so the blade is only about an inch long, makes an excellent chisel for the purpose. Sharpen the blade and you will have a chisel about 1 or 1½ inches wide, very thin, but short and stubby, which makes it strong. With this you can get down into the crack between two boards and chisel off the tongue as far as necessary. Then bore two holes in the board as close as possible to joists—see Fig. 13–13. With a keyhole saw, cut across as close to the joist as you can. It is best to cut at an angle so that the board, when re-placed, forms a wedge. The board should be removed over the space of at least three joists so when re-placed it rests directly on at least one joist. When replacing the board, first nail a cleat to the joist at each point where you sawed across. These cleats must be very solidly nailed so that when the board is replaced there will not be any springiness.

Fig. 13–13 Sometimes floor boards must be lifted. If the boards are tongued and grooved, cut off the tongue on each side with a very thin chisel, then saw across next to the joists.

Temporary openings in walls   Sometimes a temporary opening must be made in a wall so that cable can be fished around a corner. On papered walls, use a razor blade or utility knife to cut through the wall covering to form a top-hinged flap as shown in Fig. 13–14. Before lifting the section of paper, soak it with a wet rag to soften the paste. Thumbtacks will hold the lifted portion out of the way while you make a temporary opening in the drywall or plaster. When the wiring is finished, paste the paper back into place.

Fish tape   See Fig. 12–8 and related text for information about using fish tapes, which are very helpful in old work for fishing cables.

Fig. 13–14 In old work, lift the wallpaper over the area of a temporary opening, as shown here. The top of the paper serves as a hinge.

Running cable to two openings on same wall   In Fig. 13–15 cable must run from opening A to opening B, both in the same wall. Depending on the structure of the building, the cable may run in one of three ways.

Under floor boards   Running the cable as in "Route 1" in Fig. 13–15 is the simplest way. Use this route if you can easily lift the floor boards in the floor above so the cable can be dropped down from above to the locations of opening A and opening B.

Through the basement   If it is very difficult to get into the ceiling space from above, it may be possible to run the cable down through the basement as shown by "Route 2" in the same figure. If the wall is an outside wall, there will probably be an obstruction where the floor joins the wall. In most cases it is possible to bore upward

Fig. 13–15 In running cable from A to B (both on same wall) there is usually a choice of three routes for the cable. The best route depends on the structure of the building.

through this at an angle from the basement. Then push two pieces of fish tape upward through the bored holes until the ends emerge at A and B. Then by pulling at A and B, fish the opposite ends of a piece of cable upward until the ends come out at A and B. If the wall is an inside wall, there may be no partitions in the basement immediately below this wall so that it should be possible to bore straight upward. Then fish the cable upward to opening A and opening B.

Behind the baseboard   In the problem above, if it is impossible to run the cable through either the ceiling space above or the basement below, use "Route 3" by running the cable behind the baseboard along the bottom of the wall. First remove the baseboard. Then make a small opening into the wall behind the baseboard directly under A and another directly under B. Between these two openings cut a channel in the wall into which the cable can be laid. Fig. 13–16 shows the completed installation. Wherever the cable crosses a stud, cover it with a $\frac{1}{16}$-inch steel plate to protect it from future penetration by nails. Finally, replace the baseboard.

From ceiling opening to wall opening   The problem in this case is to run cable from an opening in the ceiling, labeled C in Fig. 13–17, around the corner at D, and down through the wall to opening E. It may be a very simple problem or a difficult one, depending on the construction of the house. If the ceiling joists run in the direction shown in the small inset of the same drawing, the problem is greatly simplified. If the floor above is easily lifted, it is then a simple matter to pull the

Fig. 13–16 Running cable in a trough cut in the wall behind the baseboard.

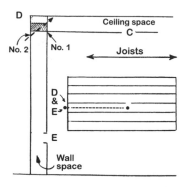

Fig. 13–17 At left, problem in running cable from opening C in ceiling, around corner at D, to E in side wall.

Fig. 13–18 At right, choice of routes for cable in Fig. 13–17 if joists run in the direction shown.

cable in at opening C, drop it down at D until it comes out at E. Even if there is an obstruction at D, as is usually the case, it is easy to bore a hole down from above after the board has been lifted.

If it is impossible to lift the floor above, the cable must be gotten around the corner at D some other way. An opening must be made into the wall. Sometimes it is made at the point marked "No. 1" by cutting through the wall, and next chiseling away part of the obstruction. Push a length of fish tape into this opening until the end shows up at C. Pull it out of C until the opposite end is at opening D. Then carefully push it down inside the wall until the end shows up at E. You then have a continuous fish tape from C around D to E. Attach the cable to the fish tape at one end, pull at the opposite end, and fish it through the wall until you have a continuous cable from C to E. Sometimes it is easier to do this from the opposite side of the wall, as at

INSTALLING SERVICE EQUIPMENT AND WIRING

point marked "No. 2," boring upward through the obstruction as shown by the dotted arrow. Then using fish tape, pull in the cable as before. After the cable has been fished, patch the wall and the job is finished.

If the opening E is not directly below point D, but is to the right or left, run the cable over to the proper point above E (if the floor board above can be removed) and drop down. If the flooring cannot be removed, drop from D down to the baseboard; behind the baseboard run over to a point below opening E, and then run upward to E.

If the joists of the ceiling run in the wrong direction, as shown in Fig. 13–18, there is again a choice of routes. If the floor above can be removed, follow "Route 1," boring holes through the joists through which the cable is to run. If the floor cannot be lifted, make an opening at point X, drop the cable down at X to the baseboard below, run it behind the baseboard, around the corner to a point below E, and from there upward to E.

No two houses are alike, so you will simply have to use your common sense in getting around obstructions. Temporary openings often have to be made, and in all cases you will probably use more material than you would for new work. The cost of a few extra feet of cable for a longer route is insignificant compared to the additional time it would take you to follow the shortest route.

INSTALLING SWITCHES, RECEPTACLES, AND OUTLETS

Replace old receptacles if they are damaged or if you wish to change the color. Options for replacing two-wire receptacles are given in this section. Much convenience can be gained from adding new switches to existing outlets and adding new outlets to existing circuits. New outlets should be only for greater accessibility—be sure they are not used to add to the total load on a circuit.

Replacing old two-wire receptacles   Where there is no grounding means in the box, there are three options for replacing two-wire receptacles: (1) two-wire receptacles, (2) three-wire receptacles grounded to the grounding electrode system or the grounding electrode conductor, or (3) GFCI-type receptacles. If you select GFCI-type receptacles, the NEC requires that any new added receptacles supplied from the GFCI receptacles must be marked "GFCI Protected" and "No Equipment Ground."

Adding switches to existing outlets   The connections in the present outlet will look a great deal like the left-hand part of Fig. 13–19. There may be more wires in the box than shown, but there will be only two wires connected to the fixture, one white and one black. In the right-hand part of Fig. 13–19 is shown the same outlet after the addition of the switch. To make the proper connections at the fixture, open the black wire splice to the fixture, thus producing two new ends of wire. These two new ends are connected to the two wires in the cable which runs to the switch. The black wire from the fixture is connected to the black wire in the new piece of cable; the black wire of the cable, which runs up to the original outlet box, is connected to the white wire re-identified as black in the new piece of cable. This is contrary to general practice, but is the one case where the NEC permits a black wire to be attached to a white. See Fig. 10–4 and the surrounding discussion for an explanation of this NEC requirement.

Fig. 13–19 Proper method of adding a single-pole switch to an existing outlet.

If a pair of 3-way switches is to be added instead of a single-pole switch, proceed as outlined above except add the two 3-way switches as discussed in connection with Fig. 10–10.

Adding outlets Fig. 13–20 illustrates an added receptacle outlet, and Fig. 13–21 illustrates an added lighting outlet. The end of the cable—marked to source—must be run to an existing outlet box which contains, in addition to the white wire, a black wire which is *always* hot. If you are in doubt as to whether one of the black wires in any box is always hot, it is a simple matter to check. Turn off the main switch. Remove the cover and the receptacle or fixture from the box from which you plan to run. Take the tape off the connections and leave the exposed ends of

Fig. 13–20 Wiring of a new baseboard receptacle.

Fig. 13–21 Wiring for a ceiling outlet.

wire sticking out of the box. Using two solderless connectors, temporarily connect each wire to one of the leads of a neon tester of the type shown in Fig. 19–2. Now turn the main switch back on. If the tester continues to light regardless of whether the switch controlling the outlet is on or off, you have found your source for the new outlet. *Be sure to turn the power off again before working on the wires.*

Alternatives to conventional outlets   If you don't wish to undertake an extensive modernizing effort, consider the following solutions to several common problems.

Fig. 13–22 This adapter permits a 3-prong grounding plug to be used with an ordinary receptacle.

Two-to-three-wire adapters   What if you live in an older house with only non-grounding two-wire receptacles, and you want to use an appliance with a three-wire cord and a three-prong plug? An alternative to replacing or adding outlets is to use a "two-to-three-wire" adapter shown in Fig. 13–22. Note that it has a green terminal lug on its side. Remove the face plate and then test to establish that the box is grounded. Using the test light shown in Fig. 19–2, insert one wire in the narrow slot in the receptacle and touch the edge of the metal box with the other wire. If the box is grounded the lamp will light. Replace the face plate, plug in the adapter, and reinstall the screw through the green lug, which is then in contact with the mounting yoke of the receptacle. This procedure is by no means the complete equivalent of having grounding receptacles in your home, but it does permit you to use appliances with three-prong plugs, and does provide some degree of protection *as long as the receptacle yoke is grounded to the grounded metal box.* But if your home is wired using metal conduit or armored cable, it would be far better to replace the two-wire receptacles with the grounding types, of course adding a grounding wire from the green terminal of the receptacle to the grounded metal switch box as discussed on pages 115–116.

Multi-outlet assemblies   Few homes have all the receptacle outlets that the occupants would like. The multi-outlet assembly in Fig. 13–23 makes outlets available at intervals of 6 to 24 inches as desired. It consists of a metal channel with

Fig. 13–23 Multi-outlet assembly provides closely spaced multiple outlets.

wires and receptacles already installed. In living rooms the channel can be installed directly above the baseboard with molding added above the channel, so the whole assembly appears to be part of the baseboard. In a kitchen it is installed above the counter at a convenient height for appliances. Similar surface raceways are available in nonmetallic form. They are easily installed, light in weight, and come with a complete line of fittings, switches, and receptacles.

Handy boxes If exposed surface wiring is acceptable, you can use the "handy" utility boxes shown in Fig. 9–5, using any appropriate wiring method.

Extension rings Where the new wiring may be permanently exposed, as in basements, it is sometimes convenient to use an extension ring, which is like an outlet box without a back. Remove the fixture of the existing outlet, mount the extension ring to the flush box, then run cable or conduit for the new run from the extension ring. Replace the fixture on top of the ring. See Fig. 13–24.

Fig. 13–24 Using extension rings makes it easy to add to existing outlets in locations where the new wiring may be permanently exposed.

GETTING STARTED—THE BASICS
WIRES, CIRCUITS, AND GROUNDING
▶ INSTALLING SERVICE EQUIPMENT AND WIRING
SPECIAL WIRING SITUATIONS AND PROJECTS

CHAPTER 14

Appliances

WIRING REQUIREMENTS for some common appliances are explained in this chapter. Appliances are connected to the power supply either by a cord and plug or by permanent wiring. Cord-and-plug-connected appliances include those that are readily portable, such as toasters and vacuum cleaners. Some large appliances are permanently connected, and others come equipped with a cord and plug to facilitate servicing. Kitchen ranges, clothes dryers, and similar appliances are located on individual circuits. Some appliances are fixed—that is, fastened in place through plumbing connections or other installation conditions; examples are water heaters, oil burner motors, and central air conditioners.

When you are in the market for a new major electrical appliance, consider the more energy-efficient models being offered. The initial cost may be higher than for a conventional appliance, but the savings in power used over the life of the appliance will more than make up for the higher price and will help to conserve energy. To encourage the use of more efficient appliances, some power suppliers have incentive programs such as a cash payment for junking an old refrigerator, or a rebate for an approved appliance purchase.

RECEPTACLES

Receptacles are rated in amperes and volts. The volt rating indicates the following restrictions:

■ rated at 125 volts—may be used at any voltage up to but not over 125

■ rated at 250 volts—may be used only at voltages over 125 but not over 250

■ rated at 125/250 volts—may be used only for appliances that operate at 120/240 volts and that require a neutral wire running to the appliance

Fig. 14–1 on the next page shows a variety of receptacles labeled A through L. Except for A, all 125-volt and 250-volt receptacles have a third opening for the third prong on a three-prong plug for connection to the equipment grounding wire as discussed on pages 67–68.

Chapter 14   Appliances   145

The 125/250-volt receptacles I and J have a third opening for the third prong on the plug for connection to the grounded neutral wire of the circuit.

Note that a plug that fits A will also fit B or C; a plug made for B will also fit C; but a plug made specially for C will not fit A or B. All of the other configurations are non-interchangeable.

At A is shown what used to be the typical 15-amp receptacle (usually made in the duplex configuration) and is now used only for replacements. The receptacle at B is the same as A, except it has the third opening for the third prong on a three-prong plug—for connection of the equipment ground*ing* wire. At C is a similar 20-amp receptacle. At D is a 30-amp receptacle for larger 120-volt loads.

At E, F, G, and H are shown respectively 15-amp, 20-amp, 30-amp and 50-amp receptacles for loads operating at 240 volts. Note that each is provided with an opening for the equipment grounding prong on a three-prong plug. *Important:* That prong and that opening must never be used for a neutral wire if the appliance operates at 120/240 volts rather than at 240 volts.

At I and J are shown 30-amp and 50-amp receptacles for appliances operating at 120/240 volts. The 30-amp is used mostly for clothes dryers, and the 50-amp for

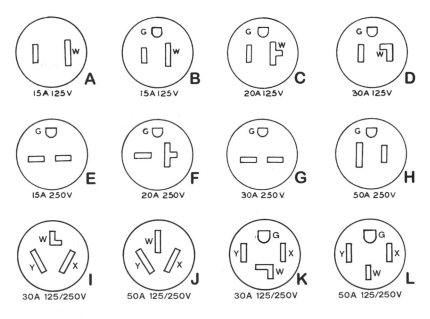

Fig. 14–1 An assortment of receptacles. Those rated at 15-amp and 20-amp have a diameter of 1.327 inches. Those rated at 30-amp and 50-amp have a diameter of 2.12 inches. In the illustrations, the opening marked G is for the equipment ground*ing* wire; the one marked W is for the white circuit wire; those marked X, Y or Z are for the hot circuit wires.

ranges on existing circuits. At K and L are shown three-pole four-wire 125/250 volt receptacles for new dryer and range circuits.

These are not the only receptacles available. There are dozens of others in two-wire, three-wire, four-wire, and even five-wire types. Besides the ordinary variety, there are others designed so the plug cannot be removed without first twisting the plug to unlock it. Some special-purpose types not used in ordinary wiring are designed so that only a plug and receptacle of the same brand will fit each other.

Fig. 14–2 Ranges are connected to 50-amp 125/250-volt receptacles of the type shown here. The pigtail is connected to the range, then plugged into the receptacle.

Receptacles come in a variety of mounting methods to fit various boxes and plates, both flush and surface types. The 50-amp receptacle is shown in both the surface-mounting type and flush-mounting type in Fig. 14–2, which also shows a typical plug (on left) with "pigtail" cord attached. These are used mostly in the wiring of electric ranges. A similar 30-amp receptacle is used for clothes dryers.

INDIVIDUAL CIRCUITS FOR APPLIANCES

The National Electrical Code (NEC) rules are quite complicated regarding when an appliance requires an individual branch circuit serving no other load. In general, you will be following the NEC rules if you provide a separate circuit for each of the following:

- Range (or separate oven or counter-mounted cooking units)
- Water heater
- Clothes dryer
- Clothes washer
- Waste disposer
- Dishwasher
- Any 120-volt permanently connected appliance rated at 12 amps (1440 watts) or more, including motors
- Any 240-volt permanently connected appliance
- Any automatically started motor such as an oil burner or pump

GROUNDING OF APPLIANCES

The following appliances must always be grounded per NEC Section 250-114(c): refrigerator, freezer, air conditioner, clothes washer, clothes dryer, dishwasher, waste disposer, personal computer, and fax machine. This grounding is especially important if the appliance is installed where a person can touch both the appliance and the ground or a grounded object. (A concrete floor, even if tiled, is considered the same as the actual earth.) For safety, always ground your appliances.

The grounding of ranges and dryers is discussed in specific sections later in this chapter. For other appliances, if they are supplied with a cord that includes a grounding wire, and a plug with a grounding blade fitting a properly installed grounding receptacle, that is all that is required.

If there is no cord and plug but the circuit wires run directly to the appliance, and if the wiring is armored cable or in conduit, check to make sure the frame of the appliance is grounded to the junction box on the appliance to which conduit or armor is anchored. But if the wiring method is by nonmetallic-sheathed cable, you must use cable with the bare grounding wire, and the bare grounding wire must be connected to the frame of the appliance.

In the case of a water heater on a farm, if the wiring is grounded to a driven ground rod but there is some buried water pipe—no matter how short—you *must* interconnect such pipe with the ground rod. This is an essential step to prevent a difference of voltage between them and to minimize danger from lightning. This is discussed in more detail in Chapter 17: *Farm Wiring*.

DISCONNECTING MEANS AND OVERCURRENT PROTECTION

Every appliance must be provided with some means of disconnecting it completely from the circuit and must be provided with overcurrent protection.

Portable appliances   The plug-and-receptacle arrangement is all that is required. The plug and receptacle must have a rating in amperes and volts at least as great as that of the appliance.

Small permanently connected appliances   If the appliance is rated at 300 watts or less (⅛ hp or less), the branch circuit overcurrent protection is sufficient. No special disconnecting means is required. Range hood fans and bathroom exhaust fans are examples of small permanently connected appliances.

Large appliances, not motor-driven   Wall-mounted electric heaters and ceiling-mounted heat lamps are examples in this category. If the appliance is connected to a circuit *also serving other loads* and if the circuit is protected by a circuit breaker, or by fuses mounted on a pullout block, no special action is needed. If the circuit is protected by plug fuses, you must install, for each appliance, a separate switch of the general type shown in Fig. 17–6. The switch need not be fused, but unfused switches are hard to find, so it is customary to use the fused kind, with one fuse for 120-volt appliances and two fuses for 240-volt appliances.

A circuit breaker may be used instead. Of course, do not install a fuse or breaker in the grounded (white) wire.

Large motor-driven appliances For motors that are *not* automatically started (dishwashers, waste disposers, etc.), proceed as if the appliance were *not* motor driven (see previous paragraph). But if the motor is automatically started, such as a water pump or an air conditioner, then provide built-in or separate motor overload protection rated at not over 125 percent of the ampere rating of the motor to protect the motor against overload or failure to start (see Chapter 16: *Motors*). While not required by the NEC, it is wise to provide an individual circuit for each automatically started motor.

WIRING OF COMMON HEAVY APPLIANCES

The NEC does not restrict the methods used for wiring heavy appliances. Use conduit or cable as you choose.

If the appliance is to be connected by cord and plug, run your cable up to the receptacle, which may be either flush-mounted or surface-mounted. The NEC demands that the receptacle be located within 6 feet of the intended location of the appliance; with good planning it should be possible to locate it even closer thus simplifying installation of the appliance.

White wire re-identified for use as black wire White wire may be used only for the grounded wire. But that wire does *not* run to any appliance operating at 240 volts. Therefore the wires running to a 240-volt load may be any color except white or green. When you use a two-wire cable to connect a 240-volt load, the cable contains one black wire and one white wire, but the white wire must not be used. What can be done? Follow NEC Section 200-7(c) instructions to re-identify the wire as described on page 103.

Provision for grounding of appliances on existing circuits NEC Section 250-140 makes an important provision that applies only to existing circuits. While many appliances must be grounded, in the case of ranges (including counter units and separate ovens) and dryers, their frames may be grounded to the neutral circuit conductor, provided it is No. 10 or heavier. See NEC Section 250-140. Moreover, for these appliances and no others, you may use service entrance cable with a bare neutral, provided it runs from the appliance directly to the service equipment.

Ranges and dryers are 120/240-volt appliances, and the neutral wire carries current in normal operation. Three-wire service-entrance cable with a bare neutral may be used in wiring 240-volt appliances such as water heaters, etc. to which the neutral does not run, provided the bare wire of the cable is used *only* as a ground*ing* wire.

Ranges In some ranges, and in all older ranges, a surface burner operates at either 120 or 240 volts, depending on whether it is turned to low, medium, or high heat. The individual burners are connected within the range in a manner that makes it impossible for the neutral wire to carry as many amperes as the two hot wires. For that reason, the wires to the range usually include a neutral that is one size smaller than the hot wires. For most ranges, two No. 6 plus a No. 8 neutral are used; for smaller ranges, two No. 8 with a No. 10 neutral are occasionally used. Surface burners on most modern ranges operate at 240 volts and have stepless control. The neutral is still needed for operating the clock, timer, oven light, etc.

Run your circuit up to the range receptacle of Fig. 14–2; this is rated at 50 amps, 125/250-volts. The range is connected to the receptacle using a pigtail cord shown in the same illustration. This also serves as the disconnecting means.

Grounding the range   The NEC requires that the frame of the range be grounded by means of a separate green grounding wire, using a four-wire cord and plug. On *existing* circuits the range is permitted to be grounded through the neutral wire, in which case a bonding strap is connected between the neutral wire and the frame of the range, and the cord and plug are three-wire. Check that the bonding strap is *not* connected in new installations where a four-wire cord is used.

Receptacle outlet for gas range   Install a receptacle outlet for a gas range. This receptacle, which can be on the small appliance circuit, is for the supply of a gas ignition system, lights, clock, and timer.

Sectional ranges   The trend is away from complete self-contained ranges, consisting of oven plus burners, toward individual units. The oven is a separate unit, installed in or on the wall. Groups of burners in a single section are installed in or on the kitchen counter where convenient. This makes for a very flexible arrangement and permits you to use imagination in laying out a custom-designed kitchen. The NEC calls such separate ovens "wall mounted ovens," and the burners "counter mounted cooking units." Here they will be referred to merely as ovens and cooking units or counter units.

Unlike self-contained ranges, ovens and cooking units are considered fastened in place. They may all be either permanently connected or cord-and-plug connected.

Two basic methods are used in the wiring of ovens and cooking units. Supplying a separate circuit for the oven and another for the cooking unit is one method. The alternate method is to install one 50-amp circuit for the oven and cooking unit combined. Any type of wiring method may be used. Regardless of the wiring method used, the frame of the oven or cooking unit must be grounded.

Where a separate circuit is installed for the oven, use wire with the ampacity required by the load. The oven will probably be rated about 4,500 watts, which at 240 volts is equivalent to about 19 amps, so No. 12 wire would be suitable. At the oven, the circuit wires may run directly to the oven, but some prefer to install a pigtail cord and a receptacle. Note that because the plug and receptacle are concealed behind an appliance that is fastened in place, the plug will not serve as the disconnecting means as it does when installing a self-contained range.

To wire the cooking units, proceed exactly as for the oven, using no smaller than No. 10 wire. This size wire has an ampacity of 30 amps and will provide a maximum of 7,200 watts, which will take care of most cooking units. Use a pigtail cord and a receptacle if you wish to make it easy to service the unit.

If you install a single circuit for oven and cooking units combined, it must be a 3-wire 50-amp circuit. Any wiring method may be used, including service-entrance cable with a bare equipment grounding conductor. The receptacles must be the 50-amp type, and may be flush receptacles installed in outlet boxes, or the surface type shown in Fig. 14–2. The circuit will be as in Fig. 14–3. The wires to the receptacles must be the same size as the circuit wires. But the wires from the receptacles to the oven or cooking unit may be smaller, per NEC Section 210-19(c) Exception 1, provided they are heavy enough for the load, not smaller than No. 12 (No. 10 if used for grounding), and not longer than necessary to service the appliance. The receptacles are not required but they may be convenient for installation. The oven or cooking unit may be connected directly to the circuit wires in the junction boxes. The NEC exception permits smaller wires between the junction box and the appliance under the same conditions specified for when receptacles are used.

Fig. 14–3 It is common to provide separate circuits for oven and counter units, but both may be placed on one circuit as shown.

Clothes dryers   An electric dryer is a 120/240-volt appliance. Wire as for a range. The NEC requires a three-pole, four-wire grounding-type 30-amp receptacle (see K of Fig. 14–1), and a four-wire pigtail cord similar to that used for a range but with smaller wires. The plug and receptacle serve as the disconnecting means.

Grounding the dryer   For existing circuits only, NEC Section 250-140 permits the frame of the dryer to be grounded to the neutral of the three wires provided it is not smaller than No. 10. Service-entrance cable with a bare neutral is permitted where it runs from the service location directly to the dryer receptacle.

Clothes washers   Washers are equipped with a cord and plug for easier servicing, so provide a 20-amp grounding receptacle on a 20-amp circuit. No separate switch is required.

Water heaters   NEC Section 210-19(a) requires that loads which are expected to be on continuously for three hours or more must not exceed 80 percent of the branch circuit rating. Dwellings and farms seldom have such continuous loads, but a domestic water heater is required to be considered in this category. This means that a 4,500-watt water heater must be on a 25-amp circuit, not a 20-amp. (4,500 watts ÷ 240 volts = 18.7 amp, but 18.7 = 93.5% of 20. The next higher standard overcurrent device rating is 25 amp, and 18.7 = 75% of 25.) This additional 25 percent of the continuous loads must be added to feeder and service calculations also, per NEC Section 220-10 and Section 230-42(a).

Look at the water heater terminal box for a marked temperature rating for the branch circuit wires. If there is no marking, the circuit wiring can be TW (or other 60°C wire), but if marked 75°C, then use a wire with an H in its designation (or HH if marked 90°C). For further information, see NEC Article 310. Where higher temperature wires are required, it is common to splice the ordinary branch circuit wires in a junction box near the water heater to short lengths of the higher temperature wire extending to the heater.

In some localities, power for heating water is sold at a reduced rate, with the heater connected to the circuit through a special electrically operated switch furnished by the power supplier. The switch connects the heater to the power line only during off-peak hours. Each day for several periods of several hours each, water cannot be heated. If your installation is of this type, do the wiring as already described, except that the wires should start from the power supplier's time switch instead of from your service equipment.

CHAPTER 15

Finishing Your Electrical Installation

ALL THE WIRING DESCRIBED in previous chapters is done as the building progresses. Installation of switches, receptacles, face plates, and fixtures—considered finishing work—usually occurs only after the walls have been papered or painted. The finishing portion of the work is a very small and usually simple part of the total job.

INSTALLING DEVICES IN BOXES

Fig. 15–1 shows how to mount a switch and its cover plate. The switch is mounted in the switch box using machine screws that come with the switch. The plate is then mounted on the switch using screws that come with the plate. Inspect any switch or receptacle and you will see that the holes for the screws that hold the device to the box are elongated rather than round. This allows the device to be mounted on a true vertical even if the box was installed crookedly (see Fig. 15–2).

Adjusting boxes Boxes must be installed so the front edges are no more than ¼ inch behind the surface of the finished wall or ceiling, provided that these consist of noncombustible materials (materials that will not burn). In all other cases boxes must be flush with the wall or ceiling surface so that when a switch, receptacle, or other device is installed inside the box, the device will protrude far enough through

Fig. 15–1 The switch or other device is first mounted in the box, and the plate is then fastened to the device.

Fig. 15–2 Elongated holes in strap allow devices to be mounted straight in a crooked box.

the opening for the plate to be attached in the proper position. If the front edge of the box is not flush with the surface and you install a switch, receptacle, or other device with its mounting yoke touching the box, the device will be so deep inside the wall that switches will be hard to operate and the job will be untidy. For this reason switches and similar devices have "plaster ears" (see Fig. 15–3) that lie on top of the wall surface and bring the device to the proper level even if the front edge of the box is behind the surface. You can easily break off the ears if they are in the way. Sometimes cheaper devices do not have plaster ears, but are provided with spacing washers—these are placed between the device and box to bring the device to the proper level so the plate will fit properly.

Fig. 15–3 Plaster ears are helpful in aligning a device flush with the wall surface.

Preventing loose connections   Here is a tip that can save you a lot of trouble: after connecting the wires to a switch or receptacle, fold the wires around to tightly hug the back of the device. Then when you push the device into place, the resistance offered by the stiffness of the wires is taken by the back of the device and not by the wiring terminals. Many loose connections, especially at receptacles, result when an acceptably tightened terminal screw is loosened as the device is pushed into place in the box.

Mounting face plates   When mounting face plates, don't draw up too tightly on the mounting screws. The common plastic plates are easily damaged. Plates for duplex receptacles have only a very narrow strip of material between the two openings; if you turn the screw too tightly you will crack this bridge and ruin the plate.

HANGING FIXTURES

It is impossible to describe all the possible ways to hang light fixtures. It all depends on the construction of the particular fixture. With the general information given here you should have no trouble because fittings to suit the particular fixture are usually supplied by the manufacturer.

Connecting the wires   On some fixtures one wire is white, the other black. More typically both are the same color, but one has a colored tracer thread woven into the covering of the wire. The white wire, or the wire with the tracer, always goes to the white wire in the box. The other wire goes either to the black wire in the box or to the switch. Connect the wires from the fixture to the wires in the box using solderless connectors.

Mounting on boxes   The National Electrical Code (NEC) requires that all fixtures be mounted on outlet boxes. Very simple small fixtures can be mounted directly on the boxes using screws supplied with the fixture (see Fig. 15–4). Somewhat larger fixtures often use a special strap supplied with the fixture. This method is shown in Figs. 15–5 and 15–6. Surface-mounted fluorescent fixtures may have a cable or conduit enter the end of the fixture, and suspended fixtures may be cord-and-plug-connected to a receptacle in a ceiling box.

Fig. 15–4 Small fixtures can be mounted directly on the outlet box.

Fig. 15–5 Sometimes a strap is first mounted on the box, then the fixture is mounted on the strap.

Fig. 15–6 Another method of mounting a fixture.

Still larger fixtures are commonly hung directly on a fixture stud mounted in the back of the outlet box, or on the fixture stud that is part of the hanger on which the box is supported. The "stem" of the fixture is threaded to fit the fixture stud. Slide the canopy down the stem. When the work at the outlet box is finished and all connections made, slide the canopy up to conceal the wiring. See Fig. 15–7.

Fig. 15–7 The fixture is supported on the fixture stud. The stem of the fixture fits the stud.

Wall brackets   For mounting a wall bracket fixture not exceeding 6 pounds in weight and 16 inches in any dimension, two No. 6 machine screws may be used. Use a switch box as if a switch or receptacle were being installed. With the wall bracket you will find a mounting strap, a short threaded nipple, and a knob (or a long heavy screw in place of the nipple and knob). Mount the strap on the box, then install the fixture on the strap. Sometimes a fixture stud is mounted in the back of the box in place of the strap. If the wall is combustible (if it can burn, as in the case of wood), the space from the box to the edges of the fixture per NEC Section 410-13 must be covered with noncombustible material such as sheet metal. (With some pan-type fixtures the noncombustible material, usually fiberglass, comes with the fixture. A pan-type fixture mounts on the surface of the ceiling and extends beyond the outlet box.)

Recessed fixtures   A portion of a recessed fixture must be installed at the rough-in stage of the work, rather than after the ceiling finish is on and painted. Most residential recessed fixtures have a connection box spaced away from the fixture to allow connection to ordinary branch-circuit wiring. However, some are marked with a temperature rating for the branch-circuit wiring which may require that you use a wire with an H or HH in its type designation (TH, THW: 75°C; THHN, XHHW: 90°C). Be sure the fixture is spaced at least ½ inch from the joists or other combustible material (except at points of support) and that it will not be blanketed by thermal insulating materials which will prevent the escape of heat (unless marked "Type I.C." meaning it will not overheat even when blanketed with thermal insulation). Do not install lamps larger than the size marked on the recessed fixture because the additional heat will damage wire insulation and could start a fire.

Track lighting   Decorative fixtures to "wash" walls with light, highlight art objects, or illuminate wall-hung paintings can be installed anywhere along a surface-mounted or flush track. Many types of fixtures are available for a great variety of decorative lighting functions. Power to the track is supplied from a ceiling outlet.

Porch fixtures   A fixture installed on the ceiling of an outdoor porch must be marked as suitable for a damp location. It is installed the same as an indoor fixture. If a wall bracket fixture is exposed to the weather, be sure it is UL labeled, marked as suitable for a wet location, and installed so that moisture cannot penetrate behind the fixture and enter the building walls.

CHAPTER 16

Electric Motors

FOR SAFETY, EFFICIENCY, AND ECONOMY, the size and type of motor you choose should be based on the requirements of the job it will be performing. This chapter discusses the work capacity, power consumption, and installation of the types of motors commonly used in homes and on farms. Special attention is given to safety factors, including grounding and protecting with overload devices.

WORK CAPACITY OF MOTORS

The capacity of a motor to do work is based on its horsepower, starting and overload capacity, and speed. Choose a motor size that is appropriate to the task.

Horsepower A motor is rated in horsepower (hp). One horsepower is defined as the work required to lift 33,000 pounds one foot (33,000 foot-pounds) in one minute. One horsepower is equal to 746 watts. Typical horsepower ratings are included in the descriptions of motor types in this chapter.

Starting capacity Motors can deliver far more power while starting than they can at full speed. The proportion varies with the type of motor; some types have starting torques four or five times greater than at full speed. Naturally the amperes consumed during the starting period are much higher than while running at full speed. In selecting a motor, it is important to consider a machine's start-up load. The motor will heat up quickly if too heavy a starting load prevents it from reaching full speed.

Overload capacity Almost any good motor will develop from 1½ to 2 times its normal horsepower for short periods after coming to full speed. Thus a 1-hp motor is usually able to deliver 1½ hp for perhaps 15 minutes, 2 hp for a minute, and usually even 3 hp for a few seconds. No motor should be deliberately overloaded continuously because overloading leads to overheating, which shortens the life of the motor. But this ability of the motor to deliver more than its rated horsepower is very convenient. For example, ½ hp may be just right for sawing lumber, but when a tough knot is fed to the saw blade the motor can instantly deliver 1½ hp and then drop back to its normal ½ hp after the knot has been sawed.

Replacing gasoline engines with electric motors   Unlike an electric motor, a gasoline engine has no overload capacity. That is why it is often possible to replace a 5-hp gasoline engine with a 3-hp electric motor. If the gasoline engine always runs smoothly and if it seldom labors and slows down, it can be replaced by an electric motor of a lower horsepower. But if the engine is always laboring at its maximum power, the motor it replaces it should be of the same horsepower as the engine because no motor will last long if it must continuously deliver more than its rated horsepower.

Speed of electric motors   The most common speed for a 60-Hz motor is theoretically 1,800 rpm. Actually the motor runs at a little over 1,750 rpm while idling, and somewhere between 1,725 and 1,750 rpm while delivering its rated horsepower. When overloaded the speed drops still more. If overloaded too much, the motor finally stalls. The speed of ordinary ac motors cannot be regulated by rheostats or switches, but there are solid-state controllers for regulating the speed of some ac single-phase motors. Special variable-speed motors are obtainable, but they are expensive special-purpose motors and will not be described here.

How temperature rise affects operation   A 10-hp motor made today isn't much bigger than a 3-hp motor made 50 years ago. This is possible because of advances made in the heat-resisting properties of insulations on the wires used to wind the motor and insulations used to separate the windings from the steel in the motor. Other advances have led to reduced air gaps and to improved magnetic properties of the laminated steel used in the pole pieces. Today's motors will run much hotter without being damaged, but motors should always be installed where they will have plenty of air for cooling.

Motors have stamped on their nameplates a "service factor" ranging from 1.00 to 1.35. Multiply the horsepower by the service factor. The answer tells you what horsepower the motor can safely deliver continuously in a location where the temperature is not over 40°C (104°F) while the motor is not running. That means the motor might develop a temperature of over 100°C (212°F), the boiling point of water, but it will not be harmed.

Most motors of 1 hp or larger have a service factor of 1.15; smaller motors have a higher service factor, some as high as 1.35. A motor works most efficiently and lasts longer if operated at its rated horsepower.

POWER CONSUMED BY MOTORS

A motor is a rotating machine that converts electrical energy to mechanical energy. A motor delivering 1 hp—746 watts—is actually consuming about 1,000 watts from the power line. The difference of 254 watts is lost as heat in the motor, friction in the bearings, the power that it takes to run the motor even when it is idling, and other factors. It is important to remember this discrepancy between watts delivered and watts consumed when selecting wire for installing a motor and when considering operating costs.

The amperage drawn from the power line depends on the horsepower delivered by the motor—whether it is overloaded or underloaded. The watts are not in proportion to the amperes, because in motors the power factor (see Glossary) must be considered. Your cost for the power is based on watts consumed, but you must provide wire size in proportion to the amperes. When the motor is first turned on it momentarily consumes several times its rated current. After it comes up to speed but is permitted to idle, delivering no load, it consumes about half its rated current. Rated current is consumed when the motor is delivering its rated horsepower, and more current is consumed if it is overloaded.

TYPES OF MOTORS

Choose the type of motor that is suitable for the job and the available power source. Described here are common ac motors. Features to consider are size, starting capacity, initial cost, operating cost, and ease of maintenance.

The direction of rotation can be changed on ac motors. On a repulsion-induction motor, shift the position of the brushes. On other types of ac motors, reverse two of the wires coming from the inside of the motor. If the motor must be reversed often, a special reversing switch can be installed for the purpose.

Single-phase motors   The motors most commonly found in homes and on farms are called single-phase motors because they operate on the usual 120/240-volt single-phase current. They are not usually available in sizes larger than 7½ hp although a few larger ones are made. The three types described below operate only on single-phase ac.

Larger single-phase motors are often dual-voltage, meaning they are designed to be operated at either of two voltages, for example 120 or 240 volts. The motor has four leads. Connected one way the motor operates at 120 volts; connected the other way it operates at 240 volts. See Fig. 16–1.

If there is a choice, always operate your motor at the higher voltage. At 240 volts it will consume only half as many amperes as at 120 volts. With any given wire size the voltage drop will be only one-fourth as great (measured as a percentage) on the higher voltage than it would be on the lower voltage.

Fig. 16–1 By reconnecting the leads, the motor can be used on either 120 or 240 volts.

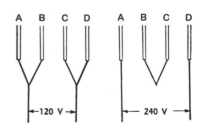

Split-phase motors   This is a simple type of motor, which makes it relatively trouble-free; there are no brushes, no commutator. It is available only in sizes of ⅓ hp and smaller. It draws a very heavy amperage while starting. Once up to full speed, the split-phase motor develops just as much power as any other type of motor, but it is not able to start heavy loads. Therefore do not use it to drive any machine that is hard to start, such as a deep-well pump, or an air compressor that has to start against compression. Use it on any machine that is easy to start, or on one where the load is thrown on after the machine is up to full speed. It is entirely suitable for washing machines, grinders, saws and lathes, and general utility use.

Capacitor motors   This is similar to the split-phase type, with the addition of a capacitor or a condenser that enables it to start much harder loads. There are several grades of capacitor-type motors available, ranging from the home-workshop type which starts loads from 1½ to 2 times as heavy as the split-phase, to the heavy-duty type which will start almost any type of load. Capacitor motors usually are more efficient than split-phase, using fewer watts per horsepower. The amperage consumed while starting is usually less than half that of the split-phase type. Capacitor motors are commonly used only in sizes up to 10 hp.

Repulsion-induction motors   This type of motor, properly called the "repulsion start, induction run" motor, is commonly called a repulsion-induction or R-I motor. Available in sizes up to 10 hp, it has a very high starting capacity and should be used for heavier jobs—it will "break loose" almost any kind of hard-starting machine. The starting current is the lowest of all the single-phase types of motors.

Three-phase motors   These motors are the simplest and most trouble-free type made and, as the name implies, operate only on three-phase ac. Three-phase motors in sizes ½ hp and larger cost less than any other type, so use them if you have three-phase current available. Do not assume because you have a three-wire service that you have three-phase current; more likely you have three-wire 120/240-volt single-phase current. See discussion on page 23. If in doubt, consult your power supplier.

LARGE MOTORS ON FARMS

On farms, motors of 10 hp, 25 hp and even larger are required. But many farms have only a single-phase three-wire 120/240-volt service. That provides only two high voltage lines to the farm and only one transformer. Single-phase motors 5 hp and larger require an unusually high number of amperes while starting, and the line and transformer often are too small to start such a motor.

Before buying even a 5-hp single-phase motor, check with your power supplier to see whether the line and the transformer serving your farm are large enough to operate such a motor. If you operate the motor only a comparatively few hours per year, your power supplier will object to installing a heavier line and transformer.

In a few localities, at least some of the farms are served by a three-phase line, providing three transformers and three wires to the farm. If you are fortunate enough to have three-phase service, your problems are solved. Use three-phase motors, which are simpler in construction so they cost considerably less than single-phase motors and rarely have a service problem. *Note:* If you have three-phase service, you will have three-phase power at 240 volts available in addition to the usual 120/240-volt single-phase for lighting, appliances and other small loads.

PHASE CONVERTERS

What can you do if you need larger motors but have only the usual single-phase 120/240-volt service? A phase converter permits three-phase motors to be operated on single-phase lines. The phase converter changes the single-phase power into a sort of modified three-phase power that will operate ordinary three-phase motors and at the same time greatly reduces the number of amperes required while starting. In other words, when operating a three-phase motor with the help of a phase converter, the same single-phase line and transformer that would barely start a 5-hp single-phase motor will start a 7½-hp or possibly even 10-hp three-phase motor, and a line and transformer that would handle a 10-hp single-phase motor (if such a motor could be found) would probably handle a 15-hp or 20-hp three-phase motor. The National Electrical Code (NEC) requirements for phase converters are in Article 455. Phase converters are expensive but their cost is partially offset by the lower cost of three-phase motors and the increased labor efficiency gained when larger machinery can be used.

Two types of converters   The *static type* of phase converter has no moving parts except relays. It must be matched in size and type with the one particular motor to be used with it; generally, there must be one converter for each motor. The *rotating type* of converter looks like a motor, but can't be used as a motor. Two 240-volt single-phase wires run into the converter; three 3-phase wires run out of it. Usually several motors can be used at the same time. The total horsepower of all the motors in operation at the same time can be at least double the horsepower rating of the converter. Thus, if you buy a converter rated at 15 hp, you can use any number of three-phase motors totaling not over 30 to 40 hp, but the largest may not be more than 15 hp—the rating of the converter. The converter must be started first, then the motors, starting with the largest and then the smaller ones.

Required horsepower ratings   Some words of caution are in order. A three-phase motor of any given horsepower rating will not start as heavy a load when operated from a phase converter as it will when operated from a true three-phase line. For that reason, it is often necessary to use a motor one size larger than is necessary for the running load. This does not significantly increase the power required to run the motor once it is started. The converter must have a horsepower rating at least as large as that of the largest motor.

The voltage delivered by the converter varies with the load on it. If no motor is connected to the converter, the three-phase voltage supplied by it is much higher than the input voltage of 240 volts. Do not run the converter for significant periods without operating motors at the same time or it will be damaged by its own high voltage. Do not operate only a small motor from a converter rated at a much higher horsepower because the high voltage will damage the motor or reduce its life. It is good practice to make sure the total horsepower of all the motors operating at one time is at least half the horsepower rating of the converter.

Check with your power supplier before buying a converter; some do not favor or permit converters. If converters are permitted, the line and the transformer serving your farm must be big enough to handle all the motors you propose to use.

MOTOR CIRCUIT REQUIREMENTS

Separate circuits are recommended for individual motors that are not part of an appliance if they are of more than ⅛ hp, whether directly connected or belt driven. Use Table 5–2, adapted from NEC Table 430–148, to determine the motor's amperage. Every motor must have a disconnecting means, a controller to start and stop it, short-circuit and ground-fault protection, and motor overload protection in case of overload or failure to start. Several of these are often combined.

Disconnecting means　If the motor is portable, the plug on the cord is sufficient. If the motor circuit starts from a circuit breaker in the panelboard, that is sufficient if the circuit breaker is within sight and not more than 50 feet from the controller, and the controller is also within sight from and not more than 50 feet from the motor and its driven machinery. The disconnecting means must show plainly whether it is on or off. If these conditions are not met, install a separate circuit breaker or a switch of the general type shown in Fig. 17–6. Use a switch with one fuse for a 120-volt motor, or with two fuses for a 240-volt motor. If the motor is 2 hp or less, the switch must be horsepower-rated or have an ampere rating at least twice the full-load ampere rating shown on the nameplate of the motor. If the motor is larger than 2 hp, the switch must have a horsepower rating equal to or larger than the horsepower rating of the motor. If a separate disconnect switch is installed, it must be installed ahead of the controller.

Controllers　A controller is any device used to start and stop a motor. It is part of the machine on refrigerators, pumps, and other equipment with automatically started motors. On manually started motors it can be a circuit breaker or switch, but is usually what is called a motor starter, as shown in Fig. 16–2. The enclosure for the starter also contains motor overload devices, which are discussed under the next heading. Controllers for motors over 2 hp must

Fig. 16–2 Typical motor starters, with built-in overload devices to prevent damage to the motor in case of overload.

have a horsepower rating equal to or more than the horsepower rating of the motor. Use the smaller starter in the illustration for fractional-horsepower motors; it has a manual switch to start and stop the motor. For bigger motors, use the larger starter shown. It has pushbuttons to start and stop the motor. Similar starters have pushbuttons in a separate case, allowing the motor to be controlled from a distance.

Motor overload devices   Motors must be protected by overload devices to prevent burnout and risk of fire resulting from extended overload.

It takes many more amperes to start a motor than to keep it running at full speed at its rated horsepower. And when a motor is overloaded, it consumes more amperes than while delivering its rated horsepower. A motor will not be damaged by current considerably larger than normal flowing through it for a short time, as at start-up or during a momentary overload. But it will burn out if more than normal current flows through it for a considerable time.

A motor overload device permits the high starting current to flow for a short time, but disconnects the motor if current due to overload (or failure to start) flows through it for a considerable time. Overload devices are permitted to be separate, but in practice they are usually included in the same enclosure with the starter.

Overload devices include "heaters" and are rated in amperes. When the starter is installed, select an overload device heater on the basis of the full-load ampere rating on the nameplate of the motor. Overload devices are often integral with (built into) the motor in the case of small motors, also with many larger motors if they are part of automatically started equipment such as air-conditioning units. Whether integral or separate, if the overload device stops the motor, correct the condition that led to the overload. Let the motor and the overload device cool off, then reset manually.

Some overload devices are automatically resetting, but they must not be used where the unexpected restarting of the motor (for example when powering a table saw) could result in injury.

Motor branch-circuit, short-circuit, and ground-fault protection   Motor overload devices, whether built into the motor or installed on the starter, are not capable of interrupting the high amperage that can arise instantaneously in case of a short circuit or ground fault that might occur in the motor circuit or in the motor. The branch circuit as well as the controller and the motor must be protected by fuses or circuit breakers against such shorts or ground faults.

The wires in a motor branch circuit must have an ampacity of at least 125 percent of the full-load motor current so they won't be damaged if the motor is overloaded. (Overload devices permit up to about 25 percent overload current for a considerable time before stopping the motor.) For motors of the kind typically installed in homes and on farms, the NEC permits a breaker to have an ampere rating up to 250 percent of the full-load current of the motor; if time-delay fuses are used, their ampere rating must not exceed 175 percent of the full-load current. If these values seem to contradict the basic requirement that overcurrent devices may not have an ampere rating larger than the ampacity of the wire being protected, bear in mind that a motor circuit is a special case in which the overcurrent device protects only against short circuits or grounds. Protection against lower values of overcurrent (overload, or failure to start) is provided by the motor overload device discussed under the previous heading. Use the smallest rating that will permit the motor to start and operate properly. Fuses that are not of the time-delay type should not be used.

INSTALLING THE MOTOR

Observe the requirement outlined above for disconnecting means, controller, and overload units. Use wire with an ampacity of at least 125 percent of the full-load current of the motor. For long circuits use wire larger than required by NEC in order to avoid excessive voltage drop. If the circuit to the motor is long, and you use wire that is too small, you may have only 100 volts at the motor during the starting period instead of nearly 120 volts. The motor may not start, or if it does it will deliver significantly less than its rated horsepower. Table 16–1 shows the wire size to use at various distances from service equipment to motor, assuming the wire has an ampacity of at least 125 percent of the full-load current shown on the nameplate of the motor. All distances are one-way. The table is for single-phase motors only.

Table 16–1 ALLOWABLE DISTANCE (IN FEET) FROM SERVICE EQUIPMENT TO SINGLE-PHASE MOTOR FOR DIFFERENT WIRE SIZES

Hp		No. 14	No. 12	No. 10	No. 8	No. 6	No. 4	No. 2	No. 1/0	No. 3/0
1/4	120 volts	55	90	140	225	360	575	900	1,500	2,300
1/3		45	75	115	180	300	450	725	1,200	1,800
1/2		35	55	85	140	220	350	550	850	1,400
3/4			40	60	100	150	250	400	600	1,000
1			35	50	85	130	200	325	525	850
1/4	240 volts	220	360	560	900	1,450	2,300	3,600		
1/3		180	300	460	720	1,200	1,600	2,900		
1/2		140	220	340	560	875	1,400	2,200		
3/4		100	160	240	400	600	1,000	1,600	2,400	
1		85	140	200	340	525	800	1,300	2,100	
1 1/2		70	110	160	280	400	675	1,100	1,700	
2		60	90	140	230	350	550	900	1,400	2,200
3				100	160	250	400	650	1,000	1,600
5					100	160	250	400	650	1,000
7 1/2						110	175	275	450	700

The wires shown in the table are compromises. The voltage drop while starting (when the amperage is several times greater than when the motor is running) will be from 3 to 7 1/2 percent depending on the type of motor, how hard the load is to start, and other factors. While running, the drop will be about 1 1/2 percent. This is less than normally recommended for non-motor loads and is the result of wire sizes large enough to keep the drop within reasonable limits while starting.

CHAPTER 17

Farms

FARM HOUSES HAVE ABOUT THE SAME power requirements as city houses and may be wired as any other house. It is the rest of the farm that makes up the main part of the load. This includes all sorts of motor-driven machinery: water pumps and water heaters; irrigation equipment; milking machines, milk coolers, sterilizers, and similar dairy equipment; extra lighting for forced egg production, incubators and brooders; crop drying equipment; heaters to keep water for livestock or poultry from freezing, and so on.

To provide enough power to operate such equipment, you must use large wires to carry the heavy amperages without undue voltage drop. The large loads require heavy service equipment and many circuits. This chapter looks at the wiring considerations encountered in providing the power needed to carry on the business of farming. Problems with large motors on farms are discussed on pages 160–161.

Wiring issues and procedures specific to farming are collected in this chapter for convenient reference, but it is important to study the entire book in order to gain understanding of the principles involved.

PLANNING YOUR FARM WIRING INSTALLATION

On practically all farms today, the power supplier's wires end on a pole in the farm yard. On the pole are the meter and often a switch or circuit breaker to disconnect the entire installation. The wires are grounded at the pole. From the top of the pole, sets of wires run to the house, to the barn, and to the other buildings to be served. At each building there is a service entrance without a meter. The type of cable you use will depend on the purpose of the building.

Locating the meter pole   Why is there a pole? Why not run the wires to the house and from there to the other buildings as was done when farms were first being wired? Because that would lead to very large wires to carry the total load involved; very large service equipment in the house; expensive wiring to avoid voltage drop—which is wasted power; a cluttered farm yard, and many other complications that can be avoided by installing a meter pole.

It is important to locate the pole as close as practical to the buildings where the greatest amount of power will be used per year; on farms, the house rarely consumes the greatest total. When the pole is located so that the largest wires are the shortest wires, you will find it relatively simple to solve voltage-drop problems without using wires larger than would otherwise be necessary for the number of amperes to be carried. Total cost is kept down because the large, expensive wires to the buildings with the big loads will be relatively short, and the smaller, less expensive wires to the buildings with the small loads will be relatively long.

Selecting cable The cable you use should have an extra bare grounding wire. The explanation for connecting it is given on pages 115-116. However, if you are using nonmetallic boxes (see next page), the bare wire needs to run only to those boxes that contain receptacles or that feed loads for which grounding is required. Buildings that house animals have special problems of humidity and corrosiveness. Only certain wiring systems may be used in such buildings.

Dangers of metallic systems Metal conduit systems are rarely used on farms. A permanently good ground is a basic requirement for conduit or armored cable installation. But on farms, poor grounds are the rule rather than the exception. In addition, buildings that house livestock usually have very high humidity, which leads to rusting of all metals including conduit. The corrosive action from the chemicals in animal waste speeds up the process. Experience has shown that when conduit systems are installed in barns and similar locations, the conduit and boxes actually rust away and ground connections sometimes disappear.

In a well-grounded and properly maintained conduit installation, an accidentally grounded wire inside conduit or inside a box leads to a blown fuse or tripped circuit breaker, indicating trouble. The problem can be corrected by finding the ground and repairing the fault. But if the conduit is rusted away near the ground, you are left with ungrounded lengths of conduit. An accidental ground inside the conduit does not lead to a blown fuse, so there is no indication of a ground. What happens if a person or an animal touches this ungrounded length of conduit with a "hot" wire accidentally grounded to the inside of it? At the minimum, an unpleasant shock; at worst, a dangerous or even fatal shock. Many electrocutions of animals can be traced to this situation. Farm animals are much less able than humans to withstand electric shocks. Many cattle have been killed by shocks that would be only uncomfortable to a person.

Nonmetallic-sheathed (NM) cable—where used Type NM nonmetallic-sheathed cable, which is described on page 32, may be used in farm houses and other buildings that do not have high humidity, but not in buildings where the humidity is normally high, and never in barns or other buildings housing livestock.

NMC and UF cable—where used Type NMC nonmetallic-sheathed corrosion resistant cable, shown in Fig. 4–5, was developed for use in buildings with damp and highly corrosive conditions resulting from housing livestock. Each conductor has its own moisture-resistant insulation, and the several conductors are embedded in a plastic jacket that is specially resistant to moisture, mildew or other fungi, and corrosion. Type NMC may not be buried in the ground, but may be used anywhere that Type NM is acceptable.

In some localities Type NMC is hard to find. In that case use Type UF cable described in this chapter on page 175. Type UF costs more than Type NMC, but may be used anywhere that Types NM and NMC may be used, and may also be buried directly in the ground.

Selecting boxes, plates, and sockets   Metal outlet boxes, plates or covers, and lighting sockets carry a risk of shock, particularly in the damp conditions frequently encountered on farms.

Nonmetallic outlet boxes assure safety   Using boxes of PVC, fiberglass, or other nonmetallic material provides insurance against shock. When metal boxes are used in barns and similar locations, they tend to rust out just like conduit. Also, when used with nonmetallic-sheathed cable, a metal box that is poorly grounded becomes "hot" when a hot wire becomes accidentally grounded inside it. Any person or animal touching the box will receive a shock that can be dangerous to humans and even more so to animals.

Fig. 17–1 Especially on farms, boxes made of insulating material are frequently used.

Nonmetallic boxes (and covers) are shown in Fig. 17–1. Use them as you would metal boxes; however, the National Electrical Code (NEC) does not require that cable be anchored with connectors to single-gang nonmetallic switch boxes if the cable is supported within 8 inches of the box measured along the cable sheath, and the sheath extends into the box at least ¼ inch. At other than single-gang nonmetallic boxes, the cable must be secured to the box, as it must to all metal boxes. (Anchoring with connectors is discussed on pages 95–96.)

Other nonmetallic boxes are shown in Fig. 17–2. Boxes with hubs, like the one at left, are for surface wiring, and with the proper covers they are weatherproof. Boxes with knockouts can be used either concealed or exposed.

Fig. 17–2 PVC boxes with integral hubs and with standard knockouts.

Nonmetallic face plates or covers are recommended when using nonmetallic boxes. If metal face plates are used, they must be grounded, which is more or less automatic with receptacles, since the mounting strap is usually connected to the green grounding terminal. Toggle switches require a special switch having a grounding terminal on the mounting strap to which the grounding wire in the nonmetallic-sheathed cable must be connected.

Choose porcelain sockets   Do not use brass-shell lamp sockets in farm buildings. To protect against shock, choose porcelain sockets. Switch-controlled lighting is preferred for safety and convenience, but if you use any lights with pull-chain control, be sure there is an insulator in the pull chain, or at least a piece of cord on the end of the chain as a safety precaution in case of accidental grounds.

Outdoor switches and receptacles   Ordinary receptacles and switches are designed for indoor use. If installed outdoors without further protection they would be very unsafe. But they can be used outdoors if installed in special waterproof boxes of cast iron or aluminum with threaded hubs instead of knockouts. Outdoor receptacles must have weatherproof covers.

Boxes for outdoor switches and receptacles   Boxes for installing outdoor receptacles are made with a variety of threaded-hub arrangements instead of knockouts. The boxes have letter designations for the different numbers and locations of the hubs. Examples are Type FS with one hub at the top and Type FSC with a hub at the top and another at the bottom of the box. Ordinary 15-amp or 20-amp toggle switches or GFCI-protected receptacles up to 50 amps may be installed in the boxes, using the proper weatherproof (gasketed) cover. If using Type UF cable (discussed on page 175) outdoors, use a weatherproof connector similar to that shown in Fig 17–3, and be sure the cable is marked "Sunlight Resistant."

Fig. 17–3 Watertight connector for cable entry into threaded hub.

Covers for outdoor receptacles   The NEC in Section 410-57 specifies two types of weatherproof receptacle covers for outdoor receptacles, the choice depending on how the receptacle is to be used:

▪ For loads that are attended while being used (such as portable tools), the enclosure may be the type with flip-covers that are weatherproof only when the lids are closed. An example of this type is shown in Fig. 17–4.

- For loads that are not attended when they are being used (such as sprinkler control systems, landscape lighting, and holiday lights), the enclosure must be weather resistant whether a plug is inserted or not. An example of a "while in use" cover is shown in Fig. 17–5.

Fig. 17–4 Flip-type receptacle covers that are weatherproof only when lids are closed. *(Leviton Manufacturing Co., Inc.)*

Fig. 17–5 Hooded receptacle cover that is weather resistant with or without plug inserted. *(TayMac Corporation)*

DETERMINING SIZE OF WIRES AND SERVICE

Wires of the proper size must run from the yard pole to individual buildings. At each building there must be a service entrance as discussed in Chapter 8: *The Service Entrance,* except the entrance will not have a meter. The service entrances must have the correct ratings, and correct ampacities must be determined for the wires running from the pole to each building and for the wires on the meter pole.

The NEC provides a specific method for determining the ampacity of the wires running from the pole to any farm building with two or more circuits, and for determining the rating of the service equipment at that building. The minimum size of wires on the pole is based on the total load on all wires between pole and individual buildings.

The wire sizes determined by using the formula that follows are the minimum permissible by NEC. You would be wise to install larger sizes to allow for future expansion. No. 10 wire is acceptable for spans up to 50 feet, and No. 8 for longer spans. For very long spans, use an extra pole. If the wires are installed in northern areas on a hot summer day, remember that a copper wire 100 feet long will be a couple of inches shorter next winter when the temperature is below zero. Leave a little slack so insulators will not be pulled off buildings during winter.

It is assumed that the building will have a three-wire 120/240-volt service, so the total amperage *at 240 volts* must be determined. For motors, use the amperage shown in Table 5–2: Determining Amperage for Motors. For all other loads, start with the wattage. For lights you can determine the total watts from the lamp size you intend to use. For fluorescent lights, add about 20 percent to the wattage because the watts rating of fluorescents defines only the power consumed by the lamp itself; the ballast adds from 15 to 25 percent. For receptacles, allowing 200 watts for each will probably be on the safe side since they will not all be used at the same time. Then divide the total watts by 240 and you will have the amperage at 240 volts.

Note: Suppose in a building you have six 120-volt 15-amp circuits for lights and receptacles—theoretically making a total of 90 amps (6 × 15) at 120 volts, or 45 amps at 240 volts. But the circuits won't all be loaded to capacity at the same time, so instead of the 45 amps, use the total as determined by the preceding paragraph.

Minimum service for each farm building   For each building to which wires run from the pole (except the house—see "What Size Service" in Chapter 8), first determine the amperage at 240 volts of all loads having any likelihood of operating *at the same time.* Enter the amperage at *a*, then proceed with steps *b, c, d, e* and *f*.

a Amperage at 240 volts of all connected loads likely to be operating at the same time, including motors, if any   _____ amps

b If *a* includes the *largest* motor in the building, add here 25% of the amperage of that motor (if two motors are the same size, consider one of them the largest)   _____ amps

c If *a* does *not* include the largest motor, show here 125% of the amperage of that motor   _____ amps

d Total of *a + b + c*   _____ amps

e Amperage at 240 volts of all other connected loads in the building   _____ amps

f Total of d + e   _____ amps

Using your results from above, select the appropriate case from the following choices to find the minimum size of service for each individual building.

■ If *f* is 30 or less, and if there are *not over two* circuits, use a 30-amp switch and No. 8 wire (No. 8 is NEC minimum). If there are *three or more* circuits, use a 60-amp switch and No. 6 wire.

■ If *f* is over 30 but under 60, use a 60-amp switch and No. 6 wire.

■ If *d* is less than 60 *and f* is over 60, start with *f*. Add together 100% of the first 60 amps plus 50% of the next 60 amps plus 25% of the remainder. For example, if *f* is 140, add together 60 plus 30 (50% of the next 60 amps) plus 5 (25% of the remaining 20 amps) for a total of 60 + 30 + 5 or 95 amps. Use 100-amp switch and wire with ampacity of 95 amps or more.

■ If *d* is *over 60 amps,* start with 100% of *d*. Then add 50% of the first 60 amps of *e*, plus 25% of the remainder of *e*. For example, if *d* is 75 amps and *e* is 100 amps, start with the 75 of *d*, add 30 (50% of the first 60 of *e*), and add 10 (25% of the remaining 40 of *e*) for a total of 75 + 30 + 10 or 115 amps. Use a switch or breaker of not less than 115-amp rating and wire with corresponding ampacity.

Calculating minimum size of wires on pole   The calculations above determined the size of the service to each building. Now let us determine the size of wires on the pole —from the top of the pole where they are connected to the power supplier's wires, down to the meter, and back up to the top of the pole where they are connected to the wires leading to individual buildings. The wires on the pole and the main disconnect must be large enough to handle all the loads calculated above. The total number you enter in No. 7 of the following calculation is the minimum rating of the switch or breaker (if used) at the pole, and the minimum ampacity of the wires on the pole.

Proceed as follows using the amperages determined on the previous page:

1 Highest of all the amperages for an individual building (excluding the house) as determined on previous page _____ amps at 100% = _____ amps

2 Second highest amperage, _____ amps at 75% = _____ amps

3 Third highest amperage, _____ amps at 65% = _____ amps

4 Total of all other buildings, _____ amps at 50% = _____ amps

5 Total all above _____ amps

6 Add the house, as figured in Chapter 8 _____ amps

7 Total of 5 + 6 _____ amps

Note 1 If two or more buildings have the same function, consider them as one building for the purpose above. For example, if there are two brooder houses requiring 45 and 60 amps respectively, consider them as a single building requiring 105 amps. If no other building requires more than 105 amps, enter 105 amps in No. 1 above.

Note 2 In listing the amperage for any one building, the amperage to use is the calculated amperage, not the rating of the switch used. For example, if for any building you determined a minimum of 35 amps but you use a 60-amp switch (because there is no size between 30 amps and 60 amps), use 35 amps not 60 amps.

Number of branch circuits in farm buildings How many circuits should a farm building have? Install enough circuits so that the business of farming can be carried on efficiently. Skimping leads to inefficiency and higher cost. Far better to install a panelboard with space for a few extra circuit breakers (or fuses) than to skimp now and pay a much higher cost per circuit later on. If a building is to be provided mostly with lights and a few incidental receptacle outlets for miscellaneous purposes, one or two circuits may be sufficient. A machinery shed may require only one circuit; a well-equipped farm workshop may require four; a dairy barn with water heaters, milking machines, milk coolers and related equipment may require a dozen circuits and large service-entrance wires in proportion to the load.

The service-entrance switch may then be anything from the simple 30-amp switch of the general type shown in Fig. 17–6 (if there are not more than two circuits) to a switch of the type shown in Fig. 8–5 which provides 12 plug-fuse 120-volt circuits (or a smaller number of 240-volt plug-fuse circuits, each pair of fuses serving one 240-volt circuit), plus a couple of cartridge-fuse circuits for motors, water heaters, and so on. By providing plenty of circuits, you lessen the likelihood of future overloaded circuits with blown fuses or tripped breakers and other troubles.

Fig. 17–6 A typical 30-amp switch for appliances and motors. Such switches are also available using cartridge fuses. Sizes larger than 30-amp will automatically have cartridge fuses. Use a switch with two fuses for 240 volts, or with one fuse for 120 volts.

BASIC CONSTRUCTION AT POLE

The three wires from the power supplier end at the top of the pole. The neutral is usually but not always the top wire; check with your power supplier. The neutral is grounded at the pole as described at the end of this section.

Installing wires on pole   Wires from the top of the pole down to the meter socket can be installed using either the single-stack or double-stack method. Use the one favored in your locality. Both methods are shown in Fig. 17-7. Regardless of the details of the installation, leave at least one-third of the circumference of the pole clear so that line workers and repair workers can climb the pole without trouble.

Single-stack construction method   In this method of installing the wires on the pole, the two hot wires and the neutral are run in one conduit down to the meter socket, and two more hot wires are run back in the same conduit to the top of the pole. The neutral ends at the meter socket, but at the top of the pole it is spliced to all the neutrals of the wires to the various buildings (see Fig. 17-7).

Double-stack construction method   In this method, all three wires including the neutral are brought from the top down to the meter socket in one conduit. Then all three including the neutral run back to the top in a second conduit, as shown in the inset of Fig. 17-7. In this method, the incoming neutral of the power supplier's line is not spliced to the neutrals of the wires to the buildings; the neutral wire that runs back to the top of the pole from the meter socket of course is so spliced. In this method, two lengths of three-wire service entrance cable may be used in place of two lengths of conduit.

Installing the meter socket   The meter socket is sometimes furnished by the power supplier but installed by the contractor. In other localities it is furnished by the contractor. Mount it securely about 5 or 6 feet above ground level.

If a switch or circuit breaker is also used, install it just below the meter socket. The two "hot" wires from the power line always run to the top terminals in the meter socket. If the usual switch or circuit breaker is used below the meter, the connections are as shown in the inset in Fig. 17-7. If this switch or circuit breaker is not used, the two wires C and D in Fig. 17-7 run directly to the bottom terminals of the meter socket. The neutral wire is connected to the center terminal on the meter for grounding purposes.

Insulators on poles   Near the top of the pole install insulator racks of the general type shown in Fig. 8-8. Provide one rack for the incoming power wires and one for each set of wires running from the pole to various buildings. Remember that there is significant pull on the wires in a heavy wind or under ice conditions in northern climates. Anchor the racks with heavy lag screws. Better yet, for each rack use at least one through-bolt all the way through the pole.

Installing the stack   Whether you use a single-stack construction (all wires in one conduit) or the double-stack, the general procedure is the same. At one time the conduit often ended at a point below the insulators, which then required great

Single-stack construction

Fig. 17–7 Typical meter-pole installation on a farm. Check with inspector and power supplier before using the single-stack construction. See NEC Section 230-7.

care in installing the wires to make sure that a drip-loop was provided, as shown in Fig. 8–19. The purpose of a drip-loop is to prevent water from following the wires into the service head. The NEC requirement today is to bring the top of the conduit to a point above the topmost insulator, as shown in Fig. 17–7, thus automatically solving the water problem.

At the top end of the conduit, use a service head of the general type shown in Fig. 8–21, with the right number of holes in the insulator. Run wires through the conduit; white for neutral, black or another color (but not green) for the hot wires. In practice, the switch, meter socket, the conduit with wires inside, and insulators are usually pre-assembled on the pole before the pole is erected. When the pole goes up, it is ready for wires to be installed on the insulators.

Connections at top of pole   The wires at the top of the pole will be of large size. Splice them using solderless connectors of the types shown in Figs. 4–22 and 4–23. Such connectors must be taped unless you use similar connectors with snap-on insulating covers.

Ground at pole   At one time it was common practice to run the ground wire out of the bottom of the meter socket to a ground rod at the pole. Better protection against lightning is obtained if the ground wire runs directly from the neutral at the top of the pole to the ground rod; see Fig. 17–11.

Remember that you will have a better ground if you install two rods 6 feet or more apart and connected to each other. When a ground rod is used (rather than an underground metal piping system), the ground wire is never required to be larger than No. 6. Tuck the ground wire in along the side of the conduit, securing it with the same straps that hold the conduit in place. In some localities it is stapled to the pole on the side opposite the conduit. Be sure to protect the ground (and especially the connection of ground wire to ground rod) against mechanical damage. Protect it against damage from livestock or from vehicles driven too close to the pole. The ground is a safety device—make it a *permanent* ground.

INSTALLING OVERHEAD WIRING

Required clearance of overhead service wires from grade is given in NEC Section 230-24(b). For feeder wires from the yard pole to the various buildings on the farm, the clearances are given in NEC Section 225-18 as follows:

- 10 feet above finished grade, sidewalks, or any platform or projection from which wires might be reached
- 12 feet above residential property and driveways and commercial areas not subject to truck traffic
- 18 feet over commercial or farm properties subject to truck traffic

For conductors over 600 volts, the National Electrical Safety Code, ANSI C2-1997 is referenced. Overhead transmission and distribution lines of the power supplier often abut or cross farm property. Clearances may have to be increased on the farm due to use of mobile equipment that is unusually high, such as grain augurs and conveyors. Dangerous situations can arise when working under an energized electric line with well-digging equipment, grain probes, irrigation pipe, and well pipe. Proper layout of the overhead electrical distribution system with relation to the structures and proximity to the power lines can save lives by minimizing exposure to potential shock. Consult with a representative of your local power supplier for advice on minimizing hazards.

INSTALLING UNDERGROUND WIRING

It is becoming more common for service entrances, and also for wires between buildings, to be put underground, producing neater installations. Underground wiring greatly reduces danger from lightning and eliminates the problem of long spans coming down under ice loads in northern areas.

Underground wiring uses either conduit or special cables designed for direct burial. The cables have special moisture-resistant insulation on the conductors, with a very tough outer layer. The NEC recognizes two types, which are described below. They are available in both single-conductor and multi-conductor configurations.

Fig. 17–8 Type USE cable is designed to be buried directly in the ground without further protection. It is available also as 2-wire or 3-wire cable.

USE and UF cable   The most common cable for underground wiring is Type USE (Underground Service Entrance), which has been used for many years. It is shown in Fig. 17–8. Another type, Type UF (Underground Feeder), may be used as an underground feeder, but must be *protected by fuses or breaker at the source;* therefore it must not be used as a service entrance conductor. However, on farms Type UF could be used on underground runs from pole to building if protected by fuses or breakers.

In the multi-conductor kind, Type UF is substantially the same *in appearance* as Type NMC shown in Fig. 4-5B. Nevertheless, they are two different kinds of cable. Type NMC may be used in dry, moist, or damp locations, but not for direct burial underground. Type UF may be buried directly in the ground and may also be used wherever Types NM and NMC may be used.

Minimum depth   Use common sense in installing any kind of underground cable. NEC Section 300-5 covers the minimum depth to which it must be buried. For *residential* installations the minimum depth is 12 inches, provided the cable is protected by a breaker or fuses rated at 20 amps or less, and is protected by a GFCI. If not so protected, the cable must be buried a minimum of 24 inches, which is also the minimum depth required for all nonresidential installations. If you use individual wires, keep them close together. In locations where the cable might be disturbed (as where it crosses roadways, or where it crosses cultivated areas where there may be future digging), bury it deeper and lay a board or similar protective material above the cable before filling the trench. Do not stretch the cable tightly; lay it in a sort of "snaky" fashion. To avoid damage to the conductors, screen any gravel or rocks out of the material used for backfilling the trench, or better yet use clean river sand. Where the *service conductors* are installed underground, a plastic ribbon is required to be installed 12 inches above the conductors to provide a warning against digging into the conductors in the future.

Protecting with conduit at ground level   Wherever underground cable emerges from the ground, the above-ground portion must be specially protected against physical damage. Run the cable through conduit starting below grade. Install a conduit bushing at the bottom to protect the cable against damage from possible sharp edges where the conduit was cut. Where the cable enters the conduit, provide a short vertical S-curve in each wire to help prevent damage as the earth settles or moves under the action of frost.

The top of the conduit may end at the meter socket, in which case the conduit is screwed into the threaded opening in the bottom of the socket. If it ends at a different location outdoors, install a service head of the type shown in Fig. 17–7. If it ends inside a building, terminate the conduit at a junction box as shown in Fig. 17–15, and continue with Type NMC cable from that point on. Of course, the conduit may end at the service equipment cabinet. If the conduit does not terminate in a grounded box or cabinet, the conduit must be grounded by a clamp or grounding bushing. If the conduit ends inside a building at a point no higher than the outside end of the conduit, fill the outside end with a sealing compound to prevent water from entering the building through the raceway.

INSTALLING SERVICE EQUIPMENT

At the house and each building served by wires from the yard pole there must be a service entrance similar to that described in Chapter 8, except that there is no meter to install (because the meter is at the pole).

Entrance at house   Let the service entrance cable or the conduit end at a point above the insulators; from there it runs directly to the service equipment inside the house. Where there is an underground metal water pipe, run the ground wire to it from the neutral bus in the service disconnect enclosure, and supplement the water pipe electrode with an additional electrode, probably a driven rod or pipe, bonding the two together. If there is less than 10 feet of metal water pipe underground, then the driven rod becomes the primary electrode, but the installation will look much the same, as the water piping (*and* gas piping, if any) *inside* the building must be bonded to the electrode and the neutral.

The wiring inside the house is the same as in other houses, and has already been covered in other chapters. Be sure to install plenty of circuits to allow for future expansion and to accommodate new appliances in the future. Spare circuits installed at the start are a good investment, since adding them later can be costly.

Entrance at other buildings How to determine the service entrance size for individual buildings is discussed earlier in this chapter on page 170. If there are more than two circuits, the NEC requires a 60-amp switch as a minimum. Be sure to observe NEC grounding requirements as outlined earlier in this chapter. Note that per NEC Section 225-36, *Exception,* in garages and outbuildings on *residential* property, the disconnecting means may be an ordinary switch (single-pole, three-way, or four-way) as used in buildings serving primarily a residential purpose, as distinguished from other buildings used primarily in the business of farming. Buildings serving primarily a residential purpose might include a garage or smokehouse.

All other buildings must have a disconnecting means that will open all the ungrounded wires. It may be located in or on the building, and may be either a switch or a circuit breaker, and it *must* be suitable for use as service equipment as evidenced by the UL "Service Equipment" label. If the building is served by two 120-volt wires including a grounded wire, use a single-pole switch or breaker. If served by three wires (or only the two hot wires) use a double-pole switch or breaker. How to determine the ampere rating of the switch or breaker is covered earlier in this chapter.

Tapping service wires at buildings If a building contains a substantial load, it should be served by wires direct from the pole. When two buildings are quite close to each other with neither building having a substantial load, both can be served by a single set of wires from the pole. Naturally the wires must be big enough for the combined load of both buildings. At the service insulators of the first building, make a tap and run the wires on to the second building, as shown in Fig. 17–9. At the second building, proceed as if the wires came directly from the pole.

Fig. 17–9 When two buildings are near each other, tap the service wires from the pole where they are anchored on the first building and run them to the second building.

If the second building is very small and requires only 120 volts, tap off only two wires including the grounded wire, as shown in Fig. 17–9. If the second building has a considerable load so that 120/240 volts are desirable, tap off all three wires. The wires to the next building must be of the same size as the wires from the pole. A separate service switch and ground are required at the second building.

GROUNDS—LOCATION AND INSTALLATION

Before proceeding, review the discussions of grounding in Chapter 7 beginning on page 59 and in Chapter 8 on pages 84–88.

Where grounds are required   In farm wiring, grounding is required at the house and at each building. If the service wires come from a yard pole, one ground must be installed at the pole and another ground at the house. NEC Section 250-32 outlines the requirements for other buildings. If a building contains only one circuit and does not contain equipment that must be grounded regardless of circumstances (such as a receptacle), no ground is required. For all other buildings, follow one of these two grounding methods:

- Install a ground at each building as is done at the pole and at the house,
- Or run an equipment grounding wire with the circuit wires. It must be the same size as the circuit wires, up to 30 amps. (For circuits over 30 amps, or for feeders, see NEC Table 250-122.) This grounding wire does not replace the grounded circuit wire. The grounding wire must be connected to metallic underground water piping, where available, and to any other electrodes existing at the building or to a supplemental additional electrode. Grounding wires should be as short and direct as possible. Where feeding one building from another, and where metal piping such as water pipe runs between the two buildings, the grounded conductor at the second building must be insulated from the enclosure and not connected to the water pipe or other grounding electrodes.

Installing ground rods   When underground piping is not conveniently located at every point where a ground must be installed, a ground rod can be used as a substitute. The usual form of the ground is a rod at least ½ inch in diameter with a steel core and a copper layer on the outside, driven at least 8 feet into the ground. Galvanized steel pipe is acceptable according to the NEC, but not always to the local inspector. Pipe must be at least ¾-inch trade size, and driven 8 feet into the ground. Often two rods or pipes, or longer ones, may have to be driven to secure a low-resistance ground that will serve its purpose. When two rods are used, they must be at least 6 feet apart—if closer, two rods are little better than one. Connect the two rods together using wire the same size as the ground wire.

Bonding to water pipe and other electrodes   Note that if there is less than 10 feet of metal underground water pipe in direct contact with the earth, it does not qualify as a grounding electrode. Since metal water piping inside the building must be bonded to the grounding electrode, in practice you must in every case bond together the water pipe, regardless of length, and another acceptable electrode (grounded metal building frame, concrete-encased electrode, driven rod or pipe, or buried plate).

If the ground rod is copper-coated, use a clamp made of copper or brass; if the ground is galvanized iron, use a clamp made of galvanized iron. An iron clamp is shown in Fig. 8–23; one copper type is shown in Fig. 17–10, but there are others similar to that shown in Fig. 8–23.

Fig. 17–10 Use a copper ground clamp with a copper rod.

Burying the ground rod   Grounds must be permanent, so use extreme care in their installation. The NEC requires that at least 8 feet of a driven rod be in contact with earth, so a standard 8-foot rod and the connection to it must be buried. If you want the connection to be accessible, use a longer rod and build a box or provide a shallow well around the clamp location to protect it from physical damage. The ground rod is driven about 2 feet from the pole (or building) after a trench about a foot deep from rod to pole has been dug. The top of the rod is a few inches above the bottom of the trench. The ground wire runs down the side of the pole (or building) to the bottom of the trench, then to the ground clamp on the rod. After inspection, the trench is filled in and the rod, the clamp, and the bottom end of the ground wire remain buried. See Fig. 17–11. Remember that you will have a much better ground if you install two or three interconnected rods at least 6 feet apart.

Fig. 17–11 In some localities, the top of the ground rod is below the surface of the ground. Fill the trench after inspection.

Grounds in buildings housing livestock should be installed so that seepage from animal waste does not saturate the ground around the rod. Chemical action in time eats up the wire, the clamp, and sometimes even the rod, so what was once a good ground turns out to be no ground at all.

WIRING ANIMAL BUILDINGS

As discussed at the beginning of this chapter under "NMC and UF cable—where used," Type NMC cable was developed for use in buildings housing animals. There are also special requirements for lighting and protection of motors.

Barn wiring Barns are generally very humid, especially in winter. Proper ventilation will greatly reduce this humidity, but few barns are sufficiently ventilated. The moisture will naturally collect in the coldest parts of the barn, and in winter that means the outside walls. If possible, avoid running cables on outside walls where the alternate wetting and drying might damage the wiring. However, if you use Type NMC or Type UF cable with nonmetallic boxes, the likelihood of damage is almost entirely eliminated.

It is best not to run cable along the bottoms of joists or other timbers because it might be subject to mechanical injury. Don't run it at right angles across the bottoms of joists even if running boards are used. The cable will receive far more protection if you run it along the side of a beam, then along the side of a joist to the middle of the aisle to each point where a light is to be installed, as shown in Fig. 17-12.

Fig. 17-12 Run cable along the substantial timber to prevent later damage. The bottom of the lamp should not project beyond the bottom of the timber.

It will take less cable (and lead to less voltage drop) if you run cable down the middle of the aisle through bored holes in the joists; the bored holes should not be near the extreme edge of the joist. The exact method will depend on the details of carpentry in the barn being wired. It is important to install cable so that it is protected as much as possible from damage and from excessive moisture.

Receptacle and lighting outlets Receptacle outlets should be installed where they will not readily be bumped by animals. Most barns have too few outlets. The right number will depend largely on the kind of barn being wired. Install enough receptacle outlets so that extension cords need not be used.

Do not skimp on the number of lighting outlets to be installed. The preferable number is one behind each animal stall. The minimum is one behind each pair of stalls. Do not install lighting outlets on the bottom of timbers, but preferably between joists so that the bottom of the lamp is flush with the bottom of the joist. Damage to the lamp is less likely to occur that way. Lights should be controlled by toggle switches located in protected spots and as high as practical so they cannot easily be damaged by animals. If you mount switches at elbow height you can operate them even if your hands are full. For convenience, have at least some of the light controlled by three-way or four-way switches located at various entrances.

Reflectors maximize light output When exposed lamps are used, half of the light goes downward, and the other half goes upward and strikes the ceiling. Ceilings in barns and similar buildings are usually dirty, so most of the light directed at them is absorbed rather than reflected. Use a good reflector for each lamp to redirect light downward. A 60-watt lamp with a good clean reflector usually gives as much useful light as a 100-watt lamp without the reflector. Reflectors are inexpensive; one type is shown in Fig. 17–13. Clean the reflectors regularly.

Fig. 17–13 Reflectors pay dividends. A 60-watt lamp with a reflector gives as much useful light as a 100-watt without a reflector. Keep reflectors clean.

Haymow fixtures and wiring Good illumination of the stairway or ladder to the haymow will help prevent accidents. Haymow lights should be controlled by a switch on the main floor. For convenience, install a pilot light at the switch.

The dust in a haymow is combustible and a fire can occur if an exposed lamp is broken. When a lamp breaks it burns out, but during that fraction of a second while it is burning out the filament is at an exceedingly high temperature (4,000°F) and can start a fire. For haymows it is best to use gasketed enclosures known as "vaporproof" fixtures, as shown in Fig. 17–14. Glass globes enclose the lamps. In many localities use of vaporproof fixtures or "dust-ignition-proof" fixtures is required in haymows.

Fig. 17–14 Use protected fixture in haymow wiring.

The NEC requires that nonmetallic-sheathed cable must be given special protection when it runs through a floor. In a haymow, there is always danger that a pitchfork will puncture unprotected cable and cause a short circuit. It is reasonable that inspectors require extra protection from the floor up to a point where hay will never be in contact with the cable. Occasionally an inspector will permit cable that is installed in the corner formed by wall and stud with a piece of board nailed over it for protection.

More often, the cable is run through a piece of pipe or conduit. Note that this does not constitute "conduit wiring." The cable is pulled through the conduit and continues to be cable wiring, with the conduit merely furnishing mechanical protection for the cable. But you must ream the ends of the conduit to remove any burrs that might have been formed, and then install a bushing on each end.

In some localities it is the custom to wire the entire haymow with rigid conduit or EMT; the cable then ends on the main floor. This is a safety measure protecting the wiring in the haymow against mechanical injury that might easily occur if cable is installed where exposed to hay forks. The conduit is not as likely to rust out in the haymow where it is not exposed to fumes and moisture as on the main floor. When wiring with conduit, the change-over is easily made as shown in Fig. 17–15.

Fig. 17–15 Method of changing from cable to conduit wiring.

NEC Section 250-86 requires that you ground isolated sections of conduit in most cases, and in all cases if it is over 25 feet long. That is easily done by connecting the bare grounding wire of the cable to a green grounding screw in one of the holes in the outlet box where the conduit begins.

Poultry and livestock confinement buildings Accumulation of dust, particularly dust with moisture, inside ordinary boxes has been identified as the cause of many fires in buildings where poultry, cattle, and hogs are housed. NEC Article 547, Agricultural Buildings, specifies requirements related to this safety issue. Consult your local electrical inspector when there is any question whether a particular building is subject to these requirements. The NEC says that where animals are housed, special precautions must be taken when any of the following is present: dust from litter or feed; water from cleaning operations or condensation; corrosive vapors from animal waste; or any combination of these.

Use molded plastic boxes as shown on the left in Fig. 17–2, with close-fitting gasketed covers similar to those shown in Figs. 17–4 and 17–5. Type NMC or UF cables must be supported within 8 inches of boxes, and must have water- and dust-tight entries. Use a threaded compression watertight connector similar to that shown in Fig. 17–3, making sure the opening in the neoprene grommet is the right size for your cable. Rigid nonmetallic (PVC) conduit is well-suited to the conditions of dust, moisture, and corrosive vapors. Rigid metal conduit, intermediate metal conduit, or EMT *may* be used as the wiring method where only dust and/or moisture is present, but are not recommended where corrosive vapors are present.

Motors must be totally enclosed or designed to minimize the entrance of dust, moisture, or corrosive particles. Lighting fixtures must be guarded where exposed to physical damage; they must be designed to minimize the entrance of dust and moisture; and if exposed to water they must be watertight—see Fig. 17–14.

Protecting against stray voltage One requirement of particular note is that an *equipotential plane* is required to be established in these buildings. This means that steel mesh embedded in the concrete floor, metal stanchions, feed troughs, water piping, milk piping and any other metal with which the animals may come in contact are bonded together, and are also bonded to the building grounding electrode system using a minimum No. 8 copper wire. Especially in a milking barn it is necessary to minimize any voltage gradient that might appear between two or more of these elements when contacted by the various parts of an animal's body. Keep in mind that farm animals are particularly sensitive to as little as 0.5 volts.

Poultry house lighting Special wiring is required for lighting designed to promote egg production. It is well known that hens produce more eggs during the winter if light is provided before sunrise and after sunset. Opinions vary as to the ideal length of the "day," but a 14-hour or 15-hour period seems reasonable, which means that the length of time the lights must be on varies from season to season and must be adjusted about every two weeks.

If all the lights are turned off suddenly in the evening, the hens can't or won't go to roost, but will stay where they are. It is necessary to change from bright lights to dim lights to dark. This can be done by manually operated switches, but an automatic time switch costs so little that manual switches should not be considered. The wiring for such switches is simple and wiring diagrams are furnished with the product.

Be sure that both the bright light and the dim light fall on the roosts, because if the roosts remain in darkness when the lights come on, the hens may not leave their roosts. Neither will they be able to find the roosts in the evening if the roosts are in darkness. One 40-watt or 60-watt lamp with a reflector for every 150 to 200 square feet of floor area is usually considered sufficient for the "bright" period. Normal spacing is about 10 feet apart. For the "dim" period, 10-watt lamps are suitable.

FARM SAFETY

In rural areas it is wise to take precautions against the hazards of darkness, lightning, and fire. Consider the following safety additions for your installation.

Installing yard lights Every farm will have one or more yard lights for convenience and accident prevention. Fig. 17–16 shows a very efficient and long-lasting high pressure sodium yard light, photocell-controlled to turn on automatically at dusk and off at dawn. It can be mounted on a pole or high up on a building wall. Unless automatically controlled, a yard light should be controllable from at least two points: from the house and from a barn or other building. This requires three wires as shown in Fig. 10–18. It is contrary to the NEC to feed a yard light by tapping the wires on the meter pole.

Fig. 17–16 Photo-cell controlled, high pressure sodium yard lights contribute to convenience and safety on the farm.

A properly installed yard light requires a considerable quantity of wire and other materials, especially if it is controlled from several points. Ordinary No. 14 wire may be electrically large enough, but often a larger wire must be used for mechanical strength for long overhead spans. It is practical to install low-voltage remote-control switching as discussed on page 190.

Lightning protection   With their isolated, scattered buildings and high structures such as silos, farms are particularly vulnerable to lightning strikes which can cause heavy damage to electrical installations and can start fires. It is not only direct lightning strikes that cause harm. A strike near overhead wires can induce very high voltages in buildings, damaging appliances as well as the wiring.

Installing lightning (surge) arresters   While proper grounding greatly reduces the extent of damage, properly installed lightning arresters (called surge arresters by the NEC) reduce the probability of damage to a very low level. Fig. 17–17 shows an inexpensive arrester that is about the size of a baseball. It has three leads. Connect the white to the neutral wire, the other two to the hot wires in the main service

Fig. 17–17 If you live in a lightning-prone area, a lightning arrester is a good investment.

disconnect. If the feeders from pole to buildings are quite long, install another arrester at each building; install it on the service equipment, letting the "neck" of the arrester project into the equipment cabinet through a knockout.

For service equipment using circuit breakers—and where there are two adjacent spare spaces for plug-in breakers—there is a surge arrester available shaped like a two-pole circuit breaker that can simply be plugged into the main bus, with a wire tail to connect to the neutral bus. This removes the dilemma of where to connect the wires from the arrester described in the previous paragraph, because NEC Section 110-14(a) requires that terminals for more than one conductor be identified as such. Most inspectors will permit the surge arrester leads to be added to the main wire terminals on the basis that the safety provided by the surge arrester outweighs the technical code violation of two wires on one single-wire terminal.

Bonding a lightning rod system   Some farms employ a lightning protective system that consists of lightning rods (called air terminals) on the roof and heavy conductors running down to a driven rod electrode. The grounding electrodes for both the lightning protective system and the electric service must all be bonded together.

Wiring the water pump for fire protection   Every farm will have an electric water system. Fire protection is a very important function of this system. However, in case of fire a problem can arise if the main switch is pulled and the water pump—the fire pump—stops. At the very moment when water pressure is most needed, there is no pressure.

A simple solution is permissible under NEC Section 230-72. Wire the pump in immediately after the meter, between the meter and the main disconnect switch on the pole. Operating the main switch will kill all electrical wiring on the farm except the water pump. This kind of installation requires a separate weatherproof switch or circuit breaker for the pump on the pole ahead of the main fuses or breaker. Beyond that switch or breaker the wiring is that of an ordinary circuit serving the pump. Information about circuits for large appliances including water pumps is given on page 149 in Chapter 14.

CHAPTER 18

Low-Voltage Wiring

LOW-VOLTAGE SYSTEMS DISCUSSED HERE operate at less than 100 volts, with most at well under 100 volts. Telephones and doorbells are common items that use low-voltage wiring—up to ten volts. Other low-voltage items include CATV, lighting outlets using transformers, wireless radio-controlled switches, central vacuum system controls, and some thermostats.

TELEPHONE WIRING

The dividing line between the telephone company's network and the customer's premises is called the demarcation point. Everything on the network side of this line is the property and responsibility of the telephone company. Equipment and wiring on the customer side is the responsibility of the customer, who may purchase equipment and material from the provider of his or her choice. In some places the telephone company installs the demarcation point equipment, consisting of connection and testing terminals, required grounding, and equipment for protection against lightning and voltage surges. In other cases it is done by the user. Check with your local telephone company to find out what the practice is in your location. You can choose to have the telephone company do the wiring inside your home, but the charges are typically high. Instead, you could hire an electrical contractor or a communications contractor to do the work inside your home, or you can do it yourself.

Safety precautions   If you decide to install telephone wiring yourself, several safety precautions should be observed.

■ Make sure the telephone ground is to the same electrode as the power ground.

■ Disconnect the telephone line at the demarcation point or take a telephone handset off the hook to prevent the phone from ringing. Telephone voltages can reach dangerous levels when the phone is ringing.

■ Do not install or use a telephone during a lightning storm because of the danger of shock to yourself and damage to equipment.

Telephone cable   Listed cables include Types CMP for use in air-handling ducts or plenums; CMR for risers between floors; CM and CMG for general use but not for plenums or risers; MP multipurpose for plenums, risers, and general use; and CMX for use only in dwellings and installed in raceways.

When stripping wire, keep in mind that the No. 22 to No. 24 sizes are relatively delicate so care is needed to avoid nicking the wire. It is safer to use terminals that displace the insulation rather than ones that require it to be stripped.

Wiring methods for multiple outlets   There are two general methods of wiring for several telephone outlets. One is to run everything in parallel (as are the lights in Fig. 6–7). Troubleshooting at a later date on such a system is difficult. The other method is to run a separate cable to each telephone outlet from the demarcation point. This method makes it easier to upgrade later.

Copper conductors are twisted together in pairs to minimize induction and crosstalk and to reduce interference from nearby sources such as motors. Two, three, or more pairs of conductors are run in one cable. Although a single pair is all that is necessary to operate a single telephone, it is wise to run multiple pairs to each outlet to accommodate future uses such as intercom, speed dialing, call forwarding, multiple lines, and other features at one telephone. One pair could be used for a computer modem, another for a fax machine. The standard color code for twisted pairs is green/red, black/yellow, and white/blue. When making connections, always match the colors of the wires to the color markings on the terminals of outlets and on other telephone equipment.

New construction   Consider adding two or more telephone lines in new construction so you can have separate lines for uses such as computer modems and fax machines. Also install plenty of telephone outlets whether they appear to be needed or not. For convenience, every floor should have telephone outlets, including the basement. These outlets can consist of pull wires from near the demarcation point to each outlet. The outlet needs only to be a plaster ring covered with a blank plate that can later be changed to a telephone jack when needed.

The telephone jacks that you install at the wall outlet are larger than the jacks built into the telephone bodies and handsets. Wall jacks are simple to connect. Each component will come with clear installation instructions.

Alternatively, cable can be installed if the pull-wire method is impractical. The best way to do the job is to use outlet boxes and empty electrical metallic tubing, or electrical nonmetallic tubing (a pliable, corrugated lightweight PVC raceway not covered in this book) which allows for changing the cable at some future time.

Place telephone cable at least 12 inches away from the electrical cable running through your house to minimize noise and to eliminate any possibility of the telephone wiring contacting line voltage. Don't mistake electrical cable for telephone cable, which is about the size of a drinking straw and is very flexible. The telephone cables should cross power cables at 90°. Do not share the same holes through wood members or the same stud spaces with electrical wiring.

Old work   To meet the needs of a growing family or a home office, you can install additional jacks and lines. From the demarcation point, telephone cable can be run on the surface using wall jacks for surface mounting. You may find that running it along the baseboard and around door and window openings is sufficiently inconspicuous. Place the cable so that it is not subject to physical damage. For a neat installation, install outlet boxes for telephone jacks and run the cable behind the walls or woodwork as described in Chapter 13: *Modernizing Old Wiring*. The outlet box is not required for a telephone jack, but it provides a base for fastening the face plate. There are also telephone jack face plates that provide a means of attaching the jack to the plate without using a box.

CATV WIRING

As with the telephone system, the outer conductor at the Community Antenna Television (CATV) entrance to the building, or your TV dish entrance, must be grounded to the same electrode as the power system by a No. 14 or larger copper wire. Where a separate electrode is used, it must be bonded to the power system electrode by means of a No. 6 or larger copper conductor. The coaxial cable must be UL Listed and resistant to the spread of fire. General-use CATV cables are identified by a CATV marking. CATV cables with a P suffix (CATVP) are suitable for use in ducts and plenums; with R suffix (CATVR), for use as risers between floors; with X suffix (CATVX), for use only in dwellings or installed in raceways.

DOORBELLS AND CHIMES

These necessary items are easy to install. The power is furnished by a transformer, a device that reduces 120-volt ac to a different ac voltage. For doorbells this is usually about 6 to 10 volts. Doorbell transformers have two "primary" *wire* leads which are permanently connected to any point on any 120-volt circuit that is not controlled by a switch, and two "secondary" *screw terminals* for the low voltage. The power consumed while the bell is not ringing is under one watt, despite the permanent connection to the line. When you push the button to ring the bell, you close the low-voltage circuit, and then the transformer draws about five watts from the 120-volt line.

Installation of doorbells and chimes is discussed here. For instructions on repairing a doorbell or chime that does not ring, see pages 202–203 in Chapter 19: *Troubleshooting and Repairs*.

Doorbells   For installation of a doorbell, consider the two secondary or low-voltage terminals on the transformer as the SOURCE for a circuit, the bell as the outlet, and the button as a switch. Breaking it down this way makes all these diagrams very simple. Fig. 18–1 shows the installation of one bell with a push button in one location. If a button is wanted in an additional location, install it as shown in the dotted line, in parallel with the first button.

Fig. 18–1 The wiring of a doorbell with a transformer is simple.

In most installations, a single bell is not considered enough. One choice is to install a bell for the front door and a buzzer for the back door. You will see that Fig. 18–2 is the same as Fig. 18–1 for a single bell, with the buzzer now added as shown with the dotted lines.

Fig. 18–2 This diagram shows how to add a buzzer for the back door.

Door chimes   Musical chimes are often used instead of doorbells. Models are available that sound two musical notes when the front door button is pushed, and a single note when the back door button is pushed. No matter how long the button is held down, the sound does not repeat. The wiring is the same as for the doorbell.

If you install chimes in place of an existing doorbell and find that the sound is not very loud, you will have to install a new transformer. The transformers used with doorbells usually deliver only about 6 to 10 volts, and many chimes require from 16 to 24 volts.

Because the usual transformer used for doorbells and chimes cannot deliver more than about 20 volts, nor more than about 5 watts, there is little danger of shock or fire even in case of accidental short circuit. Therefore the wire used for this low-voltage wiring does not need much insulation. Bell wire, which is insulated with only a thin layer of plastic, is commonly used (see Fig. 18–3). Attach it to the surface using insulated staples.

Fig. 18–3 Shown above are wires for connecting doorbells and chimes. Because of the low voltage involved, such wires need little insulation.

Wireless radio-controlled doorbell system An alternative to the wired-in doorbell or chimes is a radio-controlled doorbell system. The sounding unit is usually electronic, and some can be programmed to play different tunes. There will be a battery, probably at the pushbutton location, which will need replacing periodically. A convenient advantage of the radio-doorbell is that the sounding unit can be taken with you to the basement, patio, or back yard so you can hear the bell wherever you are.

LOW-VOLTAGE SWITCHES

Using a transformer provides an extremely flexible system for control of lighting outlets from any number of switch locations. The 120-volt branch-circuit wires are run to each lighting outlet, but all the switch legs are low-voltage wires—easy and economical to install.

In each outlet box where power is consumed, a very small relay (which is an electrically operated switch) is installed with one end projecting out of the outlet box through a knockout. A 24-volt transformer installed in some convenient location furnishes power to operate all the relays. Because the voltage is low and the transformer delivers only a very few watts even if short-circuited, the wiring between the transformer, the relays, and the individual switches resembles doorbell wiring. Inexpensive wire, insulated for only a low voltage, is run from the transformer to each relay, then to as many switches as you wish. The wire does not need to run through conduit nor does it have to be cable of the armored or nonmetallic-sheathed type. Run it or fish it through walls and staple as required. The switches do not need outlet boxes. Each switch is really a double pushbutton. Pushing the top end of the button on any switch turns on the light; pushing the bottom end turns it off.

Fig. 18–4 shows one of the relays, Fig. 18–5 one of the switches, and Fig. 18–6 the general type of wire used. Fig. 18–7 shows the wiring circuit.

120 volts   Low voltage

Fig. 18–4 The low-voltage (small) end of this latching relay projects out of the outlet box. Some models are solid-state (electronic) with no moving parts.

Fig. 18–5 The switches used in remote-control wiring are shown in two views above, with cover at right. The same kind of switch is used at every location. Use as many switches as you wish to control one outlet (or several outlets if you wish).

Fig. 18-6 There are only 24 volts on the control wires in remote-control wiring, so inexpensive wire is used.

Fig. 18-7 Typical remote-control circuit. One transformer is used for the entire house. The same circuit is often used for remote control yard lights.

You have a choice of procedures for installing the relays. One way is to run the low-voltage wires to each box, letting at least 6 inches project into the box through the knockout in which the relay is to be later installed. The second approach is to install the relays in the boxes first, and then connect the low-voltage wires to them later, allowing at least 6 inches slack in those wires. Unless the extra length of wire is provided, you would later find it impossible to replace any relay that proves defective. As an alternate to installing relays at the controlled outlets, relays can be ganged in a centrally located box which has separate compartments for the circuit wiring and the low-voltage wiring.

WIRELESS RADIO-CONTROLLED SWITCHES

If you have a hallway, stairway, basement, or attic where it would be convenient to have the light controlled by three-way switches, but the addition of the second switch and the wiring to it seems like too big a job, consider a radio-controlled set of three-way switches. One switch—the receiver—replaces the present single-pole switch. The other switch—the sender—is operated by a battery. It is surface-mounted, so no cutting into the walls is required. These wireless switches operate just as standard three-way switches do—the light can be turned on or off from either location—but they are much easier to install.

Switching of loads plugged into receptacles can also be done by radio-operated switches. The receiver plugs into a wall outlet, and the load plugs into the receiver. The sender can be up to 50 feet away. If you use more than one of these radio-controlled switches, be sure each operates on a different frequency so they will not interfere with each other.

CENTRAL VACUUM SYSTEMS

The motor of a central vacuum system is started by the closing of a low-voltage control circuit when an inlet is opened and the vacuum hose connected. A metal tank containing the motor and a collection bag is permanently wired in a utility room or in an attached garage. Airtight hose about 1½ inches in diameter is run to wall inlets in convenient locations. The wall inlets are gasketed so that when not in use they do not leak air into the system.

CHAPTER 19

Troubleshooting and Repairs

THIS CHAPTER EXPLAINS how to diagnose problems with existing wiring and how to make common repairs.

TROUBLESHOOTING FUSES AND CIRCUIT BREAKERS

The different types of fuses and circuit breakers are reviewed in Chapter 5 on pages 42–44. A fuse blows or a breaker trips for one of two reasons. Either something connected to the circuit is defective, thus drawing an excessive number of amperes, or there are too many lights, appliances, or motors connected at the same time, overloading the circuit.

If a fuse blows or breaker trips quickly every time a particular appliance is plugged in, especially if it makes no difference whether it is plugged into a different circuit, the appliance is defective. Often the defect is in the cord. If the appliance has a removable cord, try a different cord; if the cord is permanently attached, only careful inspection will locate the defect. Badly twisted and worn cords must be replaced, not repaired. See "Troubleshooting Cords and Plugs" in this chapter on page 200.

If a fuse blows when a motor (as in a home workshop) is turned on, remember that a motor that consumes only 6 amps while running may draw over 30 amps for a few seconds while starting. Substituting a time-delay fuse for the ordinary fuse might solve the problem. If the fuse continues to blow, suspect the motor. Check the cord. Check to see that the motor bearings have oil. (Some motors have sealed bearings never requiring oil.) Perhaps the belt is too tight, or the machine that the motor drives lacks oil, increasing the load on the motor and increasing the amperes beyond the safe point.

If fuses blow or circuit breakers trip from time to time on any one circuit for no apparent reason, it is likely that the circuit is simply overloaded. You will need to disconnect some of the load on that circuit. The wise procedure would be to install an additional circuit. Alternatively, the trouble may be in the wiring of the circuit protected by that particular fuse or breaker, and the short circuit or ground must be located and repaired.

Diagnosing blown fuses Usually by looking at a fuse you can tell if it is blown and what made it blow—an overload or a short circuit.

Plug fuses To tell if a plug fuse is blown, look through the glass window on top. In a normal fuse you will see a small link of metal, usually with a narrower portion in the middle, as shown in A of Fig. 19–1. If the fuse has blown because of an ordinary overload, it will have the appearance of B in the same illustration; the narrow part of the link is gone. If the inside of the window is spattered with metal as in C, it has blown because of a short circuit. In that case, you must investigate and correct the cause of the short circuit before replacing the fuse.

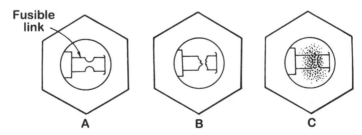

Fig. 19–1 Plug fuse: A—normal; B—blown because of overload; C—blown due to short circuit.

Time-delay fuses If your fuses are of the time-delay type shown in Fig. 5–3, note the coiled spring in a normal fuse. If such a fuse blows, look at the spring. If it is still extended, the fuse has blown because of short circuit. If the spring is contracted so that its coils touch each other, it has blown because of overload.

Cartridge fuses These rarely show any outward sign of having blown. Test in place with a neon tester, or remove the fuse from the fuseholder and test it with a continuity tester.

Fig. 19–2 An inexpensive neon test light can be used on either 120-volt or 240-volt circuits and is also handy for testing fuses. It contains a tiny neon bulb that consumes only the smallest fraction of a watt and lasts many thousands of hours. *(A. W. Sperry Instruments Inc.)*

The *neon tester* in Fig. 19–2 can be used to test fuses in their holders if you can get at the terminals of the fuseholder. With the circuit on, touch the two leads of the tester to the two terminals of the fuseholder. If the light in the tester glows, the fuse is blown. If it does not glow, the fuse is not blown. If the cartridge fuses are installed in a fuse-holding pullout block, note that the pullout block usually has small holes through which the leads of the tester can be inserted without removing the block from its holder.

A *continuity tester* is pictured in Fig. 11–7. To use this type of tester you must first remove the fuse from the fuseholder using a plastic fuse puller. If the fuses are in a pullout block they can safely be removed from the pullout block without the need for a fuse puller. Then touch the tester's prod to the metal ferrule or blade at one end of the fuse, and touch the alligator clip to the other end. If the lamp lights, the fuse is good. If the lamp does not light, the fuse is blown.

Resetting circuit breakers Most breakers are reset after tripping as shown in Fig. 5–6. A few brands are made so the handle returns to the off position when the breaker trips, and it is only necessary to move it back to the on position (instead of forcing it beyond off before returning it to on, as in most brands).

Labeling fuses and circuit breakers Labeling will be especially useful if your home is provided with fuses because when a fuse blows it is sometimes difficult to determine exactly which one has blown. If your fuses are labeled, when a light or receptacle is apparently without power you can go quickly to the right fuse and replace it. When a breaker trips it is easy to see which one has tripped, but labels are useful when you want to turn off a particular circuit for a reason such as installing a new switch.

First mark each fuseholder or breaker with a number. Then proceed by trial and error to determine which outlets are controlled by which fuse or breaker. Remove one fuse, or turn off one breaker, and find out exactly which outlets are then dead.

When you have determined which outlets are controlled by each fuse or breaker, make up a list and secure it to the inside cover of your fuse or breaker cabinet. The list might read something like this:

1　Main fuses (pullout block)
2　Large bedroom and hall lights and receptacles
3　Small bedroom lights and receptacles, and bathroom lights
4　Bathroom receptacles
5　Receptacles in living room and ceiling lights in kitchen and dining room
6　Bottom halves of all kitchen receptacles, and all receptacles in dining room
7　Top halves of all kitchen receptacles
8　All basement lights and receptacles
9　Range (pullout block)
10　Clothes dryer (pullout block)
11　Water heater (pullout block)
12　Clothes washer
13,14　Spares for future new circuits

UPDATING ALUMINUM WIRING

"Old technology" (pre-1972) aluminum wire in the branch circuit sizes, No. 10 and No. 12, had a coefficient of expansion that caused it to expand and flow away from beneath a terminal screw when heated by current running through it. Then, when it cooled, the connection would be loose. To compound matters, oxygen would get to the aluminum at the loose terminal, and aluminum oxide, a non-conductor, would form on the exposed aluminum wire. Not every termination behaved in this way, but enough did that by November 1971 the aluminum alloy used in electrical conductors was changed and the terminations on switches and receptacles were redesigned. Some aluminum wiring installations may need to be updated in order to prevent the possible development of loose connections, excessive overheating, and fire. If the 15-amp and 20-amp branch circuits in your dwelling were wired with aluminum conductors, a review of the section on aluminum wiring on page 31 will be helpful as you prepare to evaluate your installation. Inspect all of your receptacles and switches to see if they are marked CO/ALR. If they are not, you should replace them with devices having the CO/ALR marking, or reconnect them with copper pigtails.

In replacing the switches and receptacles, if the aluminum wire is copper-clad any kind of terminal may be used. (Terminals are discussed on pages 34–37.) Push-in terminals, shown in Fig. 4–16, may be used with No. 14 copper or copper-clad aluminum, but *not* with ordinary aluminum. Receptacles marked CO/ALR are not easy to find and may have to be special ordered.

The alternative method is to "pigtail" a short copper jumper to connect to each switch or receptacle terminal, *being sure to use splicing devices UL Listed for use with both aluminum and copper in combination.* This way the aluminum circuit wire will not depend upon the device terminals for continuity. A pigtail splice is shown in Fig. 4–15, but the twist-on splicing device shown there may not be UL Listed to connect copper and aluminum. One that is Listed is a tool-applied pressure connector that requires a trained electrician to do the work. This is not a do-it-yourself job for the average homeowner.

Consult a qualified electrical contractor if you feel your aluminum-wired house needs attention and you cannot find CO/ALR devices.

REPLACING WALL SWITCHES

Wall switches can wear out with constant use. Replacement is simple. First turn off the main switch or breaker for the building, or at least unscrew the fuse or trip the breaker that protects the circuit on which the switch is installed. Never work with hot wires. Remove the face plate over the switch, then remove the two screws holding the switch to the box. Pull the switch out of the box.

Replacing single-pole switches   If the switch you are replacing is a single-pole type, it will have only two terminal screws. Loosen the screws, remove the wires, and connect them to the two screws of your new switch. It makes no difference which wire goes to which terminal. Be sure the loop on your wires is turned clockwise around the screw (see Fig. 4–13). Reinstall the switch in the box and reinstall the face plate.

Replacing three-way switches   If the switch you are replacing is the three-way type, it will have three terminal screws. One of them will be a dark, oxidized color, much darker than the others. In disconnecting the wires, make careful note which wire runs to the dark-colored screw of the old switch; that wire must run to the dark screw on your new switch. It makes no difference to which screws the other two wires are connected.

Some switches do not have terminal screws; instead the wires are pushed into openings on the switch. To remove them, just push a screwdriver with a small blade into the slot near the openings for the wires and pull them out.

TESTING AND REPLACING WALL RECEPTACLES

To determine if a receptacle is live, insert the two leads on the neon tester shown in Fig. 19–2 into the two openings of the receptacle; if the lamp lights, the circuit is live. If the lamp doesn't light, you should consider whether the receptacle could be worn out. Its interior contacts may be bent so they no longer make good contact with the blades of a plug, in which case the receptacle should be replaced. If your receptacles are of the grounding type, as in Fig. 7–11, insert one lead of the tester into the round opening on the receptacle, and the other lead first into one of the two parallel openings, then the other. If the receptacle is properly installed, the tester should light when the second lead of the tester is inserted into the narrow slot that leads to the hot wire to the receptacle, but not when it is inserted into the wide slot connected to the grounded wire.

As in replacing a switch, turn off the main switch or breaker of the building, or at least unscrew the fuse or turn off the breaker that protects the circuit on which the receptacle is installed. Remove the face plate, remove the two screws holding the receptacle to the box, and pull the receptacle out of the box. Note that it will have two silver-colored terminal screws (called the "white" terminals) and two brass terminal screws. Like switches, some receptacles have no terminal screws. Remove the wires as described under switches. If your new receptacle has terminal screws, connect the white wires to the silver-colored screws, the others to the brass screws.

If the receptacle location is one that requires GFCI protection under the present National Electrical Code (NEC)—bathroom basin, outdoors, at kitchen counters or bar sink, garage, in crawl spaces and attics having equipment requiring servicing—the replacement must be GFCI-protected.

Replacing two-wire receptacles   If the receptacle you are replacing is the old-fashioned kind (for two-prong plugs and without a green terminal screw for a grounding wire), the replacement method depends on the type of wire used in your house. If your house is wired using metallic conduit or armored cable, replace the receptacle with a grounding type receptacle. Run a wire from the new receptacle's green terminal to the box in which it is installed, unless the receptacle is of the type specially designed to be acceptable for use without the grounding wire. If the box is not grounded, you may replace the receptacle with another nongrounding type, with a GFCI type, or with a three-wire receptacle grounded to the grounding electrode system or to the grounding electrode conductor.

In either case, note which wire(s) run to the brass terminal screws of the old receptacle, and connect them to the brass screws on the new receptacle. Connect the other wire(s) (which should be white) to the silver-colored screws, fold the wires behind the receptacle (see "Preventing loose connections" on page 154), and reinstall in the box.

If you want to use a two-wire receptacle for an appliance that has a three-prong plug, an alternative to replacing the receptacle is to use the two-to-three wire adapter pictured in Fig. 13–22 and discussed in the related text.

Replacing three-wire receptacles   If your receptacle is the grounding type for three-prong plugs, be sure the same wire that connects to the green screw of your old receptacle is connected to the green screw of your new receptacle. *Unless this is done, the receptacle is wrongly connected and can be very dangerous.*

Testing for faults in three-wire circuits   A three-wire circuit analyzer, shown in Fig. 19–3, will indicate if the three-wire circuit is correctly wired. Possible faults— open ground, open neutral, open hot, hot/ground reverse, and hot/neutral reverse— are indicated by various combinations of lights on the tester. Codes on the analyzer package will give the reason for the wiring fault so you will know how to fix the problem. The reversed connections are easily corrected. The opens may be difficult to track down, but it is very rare that a conductor in a cable is open between outlets.

The open will probably be found at a poor splice in a junction box or a poor connection at a receptacle. From the outlet where the tester shows an open, go back halfway to the panelboard and test again. If the open still shows up, you know the trouble is between that point and the panelboard. By dividing the circuit run in half and testing the remainder each time, you should be able to locate the trouble.

Fig. 19–3 A three-wire circuit analyzer is used to test for faults in three-wire circuit receptacles. *(A. W. Sperry Instruments Inc.)*

TROUBLESHOOTING LIGHT FIXTURES

Be sure to observe any maximum lamp size marking on fixtures, and relamp with the proper size. This is particularly important with recessed fixtures and with pan-type fixtures having lamps close to the ceiling. Many fires have started due to the heat from too large lamps with too high wattage.

Repairing fluorescent fixtures   The usual fixture for residential use is 4 feet long with two 40-watt tubular lamps. Some older fixtures may have a starter for each lamp. The starter will be found near the end of and behind each lamp. If the lamp doesn't light, or flickers, or glows at the ends, the first place to look is the starter. To remove a starter, rotate it 90° counter-clockwise and pull out. Replace it with a new starter that has the same wattage as marked on the old starter. If replacing the starter does not eliminate the trouble, or if the lamps are blackened at the ends, replace the lamps. The usual bi-pin lamp is removed by rotating the lamp 90° in either direction and then sliding the lamp out of the socket. Replace with the same length and wattage as marked on the old lamp.

A lamp socket that is cracked or broken must be replaced. Be sure the circuit is disconnected before working on a socket. The pan or cover over the fixture body must be removed to get at the sockets and their wiring. Carefully observe how the socket is supported and connected. Take the old one to the store to be sure of a match, and install the new socket.

If it still does not light, and you have eliminated everything but the ballast (see pages 16–17 for a discussion of the operation of fluorescent lights), you might consider replacing the whole fixture. Replacement ballasts often cost more than a new fixture. To replace a ballast, first disconnect all the wires and then loosen the supports, usually two bolts and wing nuts. Again, be sure the replacement matches both electrically and physically.

Some fluorescent desk lamps have a pushbutton switch that functions as a starter. The button must be held down for a few seconds to pre-heat the lamp. Other than this switch, the other elements are similar to those of the larger fixtures.

Repairing incandescent ceiling fixtures   If your incandescent ceiling fixture does not light, the first thing to do is to check the lamp in the fixture, and replace it if it has burned out. The next suspect is the wall switch, which may have failed. Replacing the wall switch is covered earlier in this chapter. It is possible the center contact in the lamp socket may have lost its spring due to heat, or it may have become corroded. With the power off, gently lift the center contact to restore some of the spring in it, and clean off any corrosion with a bit of sandpaper. Loose connections to the circuit wires are a possibility but not a common occurrence—an improperly installed solderless connector may have come loose.

TROUBLESHOOTING CORDS AND PLUGS

Never yank on a cord to unplug something because this puts all the strain on the fine wires inside the cord. Sooner or later some of the strands of the wire will break, and you will have a short circuit. A fuse will blow and your cord will be ruined. Instead, grasp the plug itself and pull it out of the receptacle. Unplug appliances with removable cords at the receptacle rather than at the appliance. When removing the cord from the appliance, grasp the plug itself instead of pulling on the cord.

Damaged cords   There is a serious danger of both fire and shock from a worn or damaged electrical cord. Be sure to do an occasional safety check of your extension cords, appliance cords, and cords on floor and table lamps, heating pads, etc. Signs of cord damage include a frayed or worn outer jacket or the appearance of damage to the wires inside the jacket. Using such a cord might be compared to driving a car with thoroughly worn-out tires that may blow out at any time—it is an unwise and unsafe policy. Don't try to repair the cord—replace it instead if possible, as on a lamp. For items like a heating pad where the cord cannot be replaced, discard the entire item and replace it with a new one.

Damaged plugs   If a cord is unusable only because the plug on its end has been damaged but the cord otherwise appears sound, it can be repaired by installing a new plug if the shortened cord is still adequate for its purpose. If a plug's blades are loose, bent or corroded, or if the body of the plug is cracked, replace it. Always replace a two-wire plug with a two-wire, and a three-wire with a three-wire, noting carefully that the green wire goes under the green screw, the black wire under the brass screw, and the white wire under the silver-colored screw. If your replacement two-wire plug is polarized (one blade narrower than the other) and your cord is Type SPT-2 (see Fig. 4–7), the wide blade and the grounded conductor must be connected together. But in Type SPT cord, where all of the insulation is the same color, it is difficult to know which is the grounded conductor. The grounded conductor is identified by either a raised ridge or a groove found in the insulation along one of the wires.

Be sure the replacement plug has a "dead front" with no live parts exposed after it is wired. This is the only type currently UL Listed. The old style with a separate fiber disk should not be reused. It has been discontinued because the "dead front" design is much safer.

The finely stranded wires used in flexible cords require considerable care in making terminations. Twist these strands together tightly before making a termination to prevent a loose strand from crossing where it does not belong.

Some means must be provided to prevent stress on the cord from being transferred to the terminals in the plug. This can be a cord-grip built into the plug (more likely to be found in heavy-duty plugs), a knot in the cord, or tape on the cord which will prevent a pull on the cord from being transmitted to the terminations.

REPAIRING TABLE AND FLOOR LAMPS

When the light doesn't come on, first eliminate a burned-out lamp bulb as the cause by testing with one you know is good. If the light still does not come on, the next most common failure is the switch that is built into the socket. The switch can be a push-through, a turn-knob, or a pull-chain type. Failures might also result from broken wires in the cord, bad connections, or dry and brittle insulation at the socket. Sometimes the contact, which is found at the center of the socket, becomes corroded.

Socket and switch problems To inspect the socket and wires, unplug the portable lamp and remove the shade and the harp (wire shade-holder). The socket is made up of two parts that must be separated. The socket includes the threaded lampholder and switch, and an outer brass shell which is lined with a fiber insulating sleeve. They fit into a cap that remains with the portable lamp. To separate the shell from the cap, press where the shell is marked "press" and use an easy rocking motion to take them apart.

If the fiber insulating sleeve inside the brass shell has become dry and brittle, it should be replaced. If it is necessary to replace the sleeve or the entire socket, take the old ones with you to the store to be sure of getting the right part.

If the center contact on the socket is corroded, the corrosion can be removed by using fine sandpaper. Some of the original "spring" of the center contact can be restored by gently lifting it up with a small screwdriver.

A defective switch will be revealed by the test procedure in the following paragraph.

Cord and wire problems Make a visual inspection of the cord. If there are worn spots in the insulation, or places where the cord has been tightly kinked, replace the cord. To check for broken wires in the cord use a continuity tester (see Fig. 11–7).

Test between the center contact in the lamp socket to the narrow blade on the plug, and from the threaded shell of the lampholder to the wide blade of the plug. In both cases, the test light should go on unless the switch is turned off, in which case the test to the center contact will show open (no circuit). Turn the switch and try again. If it still tests open, either the switch is faulty or there is a poor connection at the socket or the plug. Test for a broken wire in the cord by making these same tests but between the plug blades and the wires in the cord, leaving the socket and switch out of the circuit.

Also inspect the insulation of the wires of the cord at the socket. If the insulation appears dry and brittle, this section of cord should not be used and needs to be replaced. If the cord is long enough (in some portable lamps, especially those with two or more sockets, there are splices at the top of the stem or in the base), it may be possible to feed through a couple of inches and cut off the bad wire. Use the remaining good wire now at the socket to make connections to the terminal screws. If the cord is too short to pull through enough good wire, replace the cord altogether. Unless the route from the lamp base to the lamp socket is a straight shot, it would be wise to use the old cord to pull in the new. Strip about 1½ inches of insulation from all four wire ends, form staggered hooks on the wires, and tape over them taking care that the resulting splice is not too bulky. Feed in the new cord at one end and pull out the old cord at the other.

Reassembly   After performing the above steps that may be necessary to repair your lamp, the socket/brass shell must be reassembled to the cap. Be sure you hear a "click" when the two parts go together. Then replace the lamp bulb, the harp, and the shade.

Broken lamp bulb   Sometimes a lamp bulb breaks leaving the lamp base in the socket with nothing much to grip to unscrew it. To remove the base, be sure the portable lamp is unplugged or the circuit is off. Wear gloves and safety glasses. Using a cork that will just fit inside the base of the broken lamp bulb, push the cork hard into what is left of the base, and it will grab the remains so you can unscrew the lamp base from the socket by turning the cork.

TROUBLESHOOTING DOORBELLS AND CHIMES

Installation of doorbells and chimes is discussed on pages 188–190. To diagnose problems, begin by checking the various components—bell, pushbutton, transformer, and wiring—until you find the part that does not work and needs to be repaired or replaced.

Testing doorbells   To check a doorbell, start at the source and check the transformer. It has two screw terminals delivering low voltage—6 to 8 volts—for the bell, which is too low to give you a shock. Short-circuit these terminals momentarily with a short piece of wire or a screwdriver. If you see a small spark, the transformer is not defective. This spark may be very faint, so conduct this test in the dark. If you get no spark, check to see that the 120-volt circuit supplying the transformer primary is alive. If it is, the transformer has burned out and must be replaced. See Figs. 18–1 and 18–2 and related text.

If the transformer is all right, the next most likely source of trouble is the pushbutton. Remove the mounting screws of the pushbutton and disconnect it from the wires. Touch the ends of the two wires together or jumper between them with a short piece of wire. Remember that the voltage is so low that you cannot receive a shock. If the bell rings, the button is at fault. If you can reach the contact surfaces, clean them with fine sandpaper and adjust them so they make good contact but will open when the button is released. Reconnect the pushbutton and try it. If the bell still does not ring, replace the pushbutton.

If both the pushbutton and the transformer are all right, check the bell or buzzer itself. Disconnect and remove the bell or buzzer and take it to the transformer location. With a couple of short wires, connect the bell or buzzer directly to the transformer. If it rings, then the trouble is in the wiring. If it does not ring, the bell or buzzer is defective and must be replaced.

Chimes—transformer and plungers   If you have replaced a bell with chimes, it may be that the transformer was not changed. Chimes require a higher voltage— 18 volts or more—than a bell or buzzer. Install a new transformer. If the chimes still do not work, check the plunger(s) for free movement. Over time the plunger(s) get gummed up with oily dust and are not free to move when the pushbutton is pushed. Use a cotton swab and solvent to clean up the plunger(s), being careful to keep the solvent off the solenoid coil wires. Remember that when testing chimes, they will operate only as contact is made—they do not ring continuously as a bell does, but "ding" once each time the pushbutton is depressed.

Wiring between components   If the pushbutton, transformer, and bell (buzzer, chimes) all check out, then the trouble is in the wiring between them. These wires are small and subject to breakage when flexed, so check at the ends where they have been worked over during installation to see whether there is a broken wire. Due to the low voltage, the insulation on these wires is adequate electrically but not very resistant to physical damage. Look for locations where the insulation has been damaged, which you can repair with a small piece of electrician's tape. Much of the wiring is probably concealed in the walls, and if the damage is concealed there is little that can be done short of abandoning it and fishing new wires in as replacements. That may be the time to look into installing a wireless, radio-controlled doorbell system described on page 190.

It is hoped that you have found this book to be useful and that you have made safety your watchword. Your personal safety and the safety of the wiring you install are both important. Follow the guidelines discussed in this book, such as disconnecting live circuits and wearing safety glasses.

Consult your local electrical inspector before you start work, and learn which edition of the National Electrical Code is in effect in your locality.

Wiring is an exacting practice and deserves your best effort, which will be amply rewarded by a job well done.

Your questions, comments, or suggestions about the content of this book are welcome. Please send them to—

Park Publishing, Inc.
350 Main Street
P.O. Box 387
Somerset, WI 54025

Phone   1-800-841-0383

Fax   715-247-5062

E-mail   wiring@simplified.com

Glossary

Alternating current (ac) Current that periodically reverses, having alternately positive and negative values.

Ambient temperature Temperature of the medium (air, earth, water, etc.) surrounding conductors, devices, or utilization equipment.

Ampacity The safe current-carrying capacity of a wire, in amperes.

Ampere (A) Unit used to express the flow of electricity; one coulomb (the measure of quantity of electric power) per second = one ampere. Water analogy: gallons per minute.

Approved Acceptable to the inspector.

Arc Unwanted flow of electricity through an insulating medium (such as air), characterized by the emission of light and heat (as opposed to controlled arcs, such as in electric-discharge lamps).

Armored cable Two or more conductors, paper wrapped, enclosed in a flexible spiral interlocked armor of steel or aluminum, and with an aluminum bonding strip under and in contact with the armor. Informally referred to as "BX".

AWG (American Wire Gauge) Standard for measurement of wire diameters in the United States.

Ballast Auxiliary equipment used with fluorescent and electric-discharge lamps to provide starting voltage and proper operating voltage, and to limit the current.

Bonding The permanent connecting together of two or more points of non-current-carrying metallic parts of raceways, boxes, etc., so that all such parts become connected together to form a continuous path to the ground.

Branch circuit An outlet or group of outlets beyond the last fuse or circuit breaker.

Bus, busbar Conductor, usually a bar of rectangular cross section, serving as the common connection for two or more smaller conductors. Part of service equipment.

Bushing   Metallic or insulating fitting used at the end of a raceway to provide a smooth surface for wire leaving the raceway and entering an enclosure.

BX   Trade name for Type AC armored cable.

Cable   Assembly of two or more wires, as in nonmetallic-sheathed or armored cable. Individual large-size wires are also referred to as cables because they are made up of several strands of smaller wire.

Capacitance   Measure of the opposition to voltage changes in an ac circuit; exhibited by capacitors, two conductors separated by a nonconductor.

Circuit   Arrangement of conductors, devices, and utilization equipment (loads) such that current will pass through them.

Circuit breaker   Switching device, manually operable, which automatically opens a circuit at a predetermined level of overcurrent. (*See also* Fuse, Branch circuit)

Codes   Local, state, and national laws governing safe wiring practices. (*See also* National Electrical Code)

Conductor   Material, usually metal, having relatively low resistance, through which current will readily flow. Examples: wires, cables, busbars.

Conduit   Steel, aluminum, or nonmetallic raceway of circular cross section through which wires can be pulled. Protects wires from physical damage.

Current   Flow of electricity. (*See also* Alternating current, Direct current, Leakage current)

Cycle   In alternating current, the time (usually 1/60 second) during which the voltage goes from zero to + maximum, back through zero to – maximum, and back to zero again.

Demand   Kilowatts consumed during a specified time period, usually 15, 30, or 60 minutes.

Demand factor   Ratio of the maximum demand to the total connected load.

Device   Electrical system component designed to carry, but not consume, electrical energy. Examples: switch, receptacle, circuit breaker.

Direct current (dc)   Current that flows in one direction only, characterized by "poles" termed positive and negative; generally supplied by batteries, but sometimes by generators.

Disconnecting means   Equipment used to de-energize (disconnect from the source of power) an entire electrical installation, or individual circuits and appliances; may consist of circuit breaker or fused switch, or the plug and receptacle in the case of small portable appliances.

Duplex receptacle   Device consisting of a pair of receptacles for plugging in two different appliances.

Enclosure   General term for outlet box, junction box, or cabinet, whether metallic or nonmetallic, enclosing wires, devices and/or electrical equipment

Equipment grounding conductor   (*See* Grounding conductor, equipment)

Fish tape   Narrow, highly tempered steel tape with a loop at one end for pulling wire through conduit and pulling cable through walls and other spaces.

Fittings   The various standardized parts used in a wiring system to make mechanical connections. Examples: anti-short bushings, solderless connectors, conduit couplings.

Fixture   Permanently wired (not portable) lighting equipment.

Fuse   A soft metal link that melts and opens a circuit at a predetermined level of overcurrent. Fuses are enclosed in a convenient case that contains the melted metal when a fuse blows, making replacement easy.

Gang   To join individual electrical devices or boxes into a single unit. Example: To mount two devices side by side, two single boxes can be converted to one "two gang" box by removing one side of each box and bolting the boxes together.

Greenfield   A trade name for flexible metal conduit available in both steel and lightweight aluminum. Also called "flex." Similar to the armor of armored cable but the wires are pulled in after the flex is installed.

Ground   The earth or some conducting body that serves in place of the earth; a conducting connection to earth, such as a ground rod or a metal water pipe.

Ground fault   Unwanted current path to ground.

Ground-fault circuit-interrupter (GFCI)   Device intended for protecting people against shock by sensing abnormal current to ground and then interrupting the circuit. Usually rated in milliamperes (thousandths of an ampere).

Ground-fault protection for equipment (GFPE)   Device intended for protecting equipment by sensing abnormal current to ground and then interrupting the circuit. Usually rated in hundreds of amperes.

Ground potential   A point in the system at zero voltage whether carrying current or not.

Ground wire   (*See* Grounding electrode conductor)

Grounded   Connected to earth or to some conducting body that serves in place of the earth.

Grounded wire (identified conductor)   In a circuit, the wire (usually white) that normally carries current and is connected to the ground at the service equipment, and is never fused or protected by a circuit breaker or interrupted by a switch.

Grounding conductor, equipment   The grounding wire connecting non-current-carrying metal parts of equipment, raceways, and other enclosures to the grounded conductor of the system (usually the neutral) and the grounding electrode conductor at the service location.

Grounding electrode conductor (ground wire)   The wire that is bonded, within the service equipment enclosure, to: the neutral conductor; the service raceway; the service enclosure; and the equipment grounding conductor, and connects them to the grounding electrode system.

Harmonics   Odd multiples of the fundamental (60-Hz) frequency caused by nonlinear loads (e.g. 180-Hz, 300-Hz). Harmonic currents do not cancel in the neutral of three-phase four-wire systems.

Hertz (Hz)   Unit used to express frequency. One hertz = one cycle per second. What used to be called "60 cycles per second" is now called "60 Hertz."

Hickey   (1) Metal coupling for supporting a lighting fixture from a fixture stud. The center is open to allow wires to enter the fixture stem. (2) Hand tool for bending rigid metal conduit.

Hot   Designating any point in the electrical system having a voltage above ground potential. Every point should be considered to be hot until you have tested it for the presence of voltage.

Identified conductor   (*See* Grounded wire)

Insulation   Material having relatively high resistance, through which current will not readily flow when the material is applied within its voltage rating. Used to enclose the metal conductor of a wire.

Labeled   Product having an attached label, symbol, or other identifying mark indicating compliance with an appropriate standard administered by a testing organization.

Lamp   Produces visible light when an electric current passes through it. Includes incandescent (not "bulb"—that is the enclosing glass envelope), fluorescent, halogen, mercury and sodium vapor, and other types.

Leakage current   Small undesirable current which flows through insulation whenever voltage is present.

Listed   Included in a list published by a testing organization indicating that the product meets appropriate standards.

Load Anything that is connected to a branch circuit, feeder, or service and consumes electricity.

Low voltage Technically, under the National Electrical Code, any circuit under 600 volts, but more commonly describes the voltage supplying communications and signaling equipment of under 30 volts to not over 100 volts, when the "under 600" voltage is referred to as "line voltage."

Lumens Measure of the amount of light produced by a lamp. One lumen falling on a surface of one square foot produces one footcandle.

National Electrical Code (NEC) Installation code for safe electrical wiring, sponsored by the National Fire Protection Association and revised on a three-year cycle. The 1999 Code is the current edition.

Neutral The wire, usually grounded, usually having white insulation, related to two or more conductors of the circuit so that it has the same potential to each of them, and so that, with a balanced load and no harmonic currents present, it carries no current.

Ohm The unit that is used to express electrical resistance or impedance. Ohms = volts ÷ amperes.

Outlet A point on the wiring system at which electric power is used, such as at a receptacle outlet or a lighting outlet.

Overcurrent Current in excess of the rated current of equipment or the ampacity of a conductor, which may result from overload, short circuit, or ground fault.

Overcurrent protective device Circuit breaker or fuses to protect conductors and equipment against overcurrent.

Overload Operation of equipment in excess of normal, full-load current rating or of a conductor in excess of rated ampacity for a sufficient length of time to cause damage or dangerous overheating.

Panelboard A panel including buses and overcurrent devices for the control of all branch circuits and sometimes feeders to other panelboards. In dwellings the service disconnecting means and main overcurrent protection are often included in the branch circuit panelboard.

Parallel circuit Connection of two or more devices or loads across the same conductors of the circuit, current flow through each being independent of the others.

Pigtail Splicing a separate length of wire to two or more conductors to provide a tap point. (*See also* Splice, Tap)

Plug Male connector attached to a flexible cord for insertion into a receptacle; also called attachment plug and attachment plug cap.

Polarized plug Plug having one of the parallel blades wider than the other so it can be inserted into a polarized receptacle only one way, preventing accidental reversal of hot and neutral sides of circuit.

Power factor The ratio of true power (or watts) to apparent power (or volts × amperes), expressed as a percentage. Where the loads are inductive, such as motors and fluorescent lighting, the current lags the voltage resulting in a power factor of, typically, 80 to 90 percent.

Raceway Wiring enclosure, metallic or nonmetallic, usually of a type having means for pulling conductors in and out. Conduit, electrical metallic tubing, electrical nonmetallic tubing, flexible conduits, surface raceways, and wireways are common types of raceways.

Receptacle Device installed at an outlet for the connection of a single plug. (*See also* Duplex receptacle)

Resistance (R) Measure, in ohms, of opposition to current flow in a conductor, device, or load. In direct current, volts ÷ amperes = ohms resistance.

Romex Trade name for Type NM nonmetallic-sheathed cable.

Series circuit Connection of two or more devices or loads in tandem so that the current flowing through each also flows through all the others. Rarely used in residential and farm wiring.

Service entrance That portion of the system between the distribution lines of the power-serving agency and the premises wiring, including the main circuit breaker or fused switch.

Service equipment Main fixed control and cutoff of the electrical supply consisting of a circuit breaker(s) or switch(es) and fuses and their accessories and connected to the load end of service conductors in a building.

Shock Accidental flow of electric current through the body of a human or animal.

Short circuit Unwanted current flow between conductors.

Single-phase current Current produced from a single alternating source; typical residential current provided by power suppliers, using two wires (for 120-volt only) or three wires (120/240-volt).

Source Starting point of the circuit or system under consideration. For a wiring system the source is the power-serving agency's transformer, for a branch circuit it is the overcurrent device at the panelboard, and for an added outlet it is the existing outlet from which the extension is made.

Splice Two or more wires connected together mechanically and electrically to provide good conductivity between them. (*See also* Pigtail)

Stray current Undesirable current, usually through the earth, resulting either from multiple system grounds or from insulation failure. Especially hazardous to large farm animals.

Switch Device for turning electrical equipment on or off by either closing or opening the circuit to the equipment; designed for use at an amperage and voltage not higher than the limits stamped into the metal mounting yokes.

Tap The point where a conductor is connected to another, usually larger, conductor mid-run to supply a load.

Three-phase current (power) Current produced from three separate alternating forces, each in a different phase of a cycle; commonly available in cities, less often in rural areas, and typically used by factories and other establishments that use many motors; supplied through either three or four wires.

Three-phase motor Electric motor that operates only on three-phase current.

Three-way switches Switches having three terminals for wires and used in pairs to control a single light from two locations.

Three-wire circuits Circuit which is the equivalent of two 2-wire circuits sharing the grounded wire for the purpose of reducing material costs and voltage drop. Also called a multi-wire circuit.

Three-wire cord Cord containing a grounding wire in addition to the two circuit wires; the grounding wire is connected to the third prong of a three-prong plug which can be used only with a grounding-type (three-wire) receptacle. The purpose of the grounding wire is to reduce shock hazard.

Three-wire entrance (single-phase 120/240-volt service) Three wires (two hot and one grounded) run from power supplier's transformer to building served, making available both 120-volt and 240-volt power.

Three-wire receptacle Grounding-type receptacle having a third opening for the third (grounding) prong on a three-prong plug. Reduces shock hazard when used with correctly wired three-wire cord and three-prong plug sets.

Transformer Device used to raise or lower ac voltage. Volts × amperes in the coil of the primary winding equal volts × amperes in the coil of the secondary winding, less the small current necessary to magnetize the laminated steel core.

Underwriters Laboratories Inc. (UL) The qualified testing laboratory that lists and labels electrical wiring materials and utilization equipment when they meet nationally recognized standards.

Volt (V) Unit used to express the electrical force that causes current to flow. Water analogy: pounds per square inch of water pressure.

Voltage drop   Loss of voltage in a wire between the source and the end, caused by the resistance of the wire and the length of the run.

Watt (W)   The unit used to express electrical power. In direct current, one volt × one ampere = one watt.

Wire   Conductor through which electric current flows.

Index

National Electrical Code® citations (in italics within square brackets) to specific NEC Articles, Exceptions, Sections, and Tables are given under some topics. These citations reflect references appearing in the text of *Wiring Simplified,* or in some cases they direct the reader to a broader discussion of a topic in the NEC than appears in this book.

Index

Index

About the authors

The late **H. P. Richter** wrote the first edition of *Wiring Simplified* in 1932. It was one of the first how-to books on wiring for general readers. It was followed by *Practical Electrical Wiring* in 1939. With the publication of each succeeding edition of these books, the author came to be regarded more and more widely as both the do-it-yourselfer's electrician and the electrician's electrician. He had the rare ability to write for both the amateur and the professional, and his books continue today in updated versions.

W. Creighton Schwan, registered professional engineer and electrical consultant, is a recognized authority on wiring standards and practices. He has been the coauthor of *Wiring Simplified* and *Practical Electrical Wiring* and regularly contributes a column, "Code Comments," to the magazine *Electrical Contractor*. His background includes positions as supervising electrical inspector in Alameda County, California; electrical safety engineer for the California Division of Industrial Safety; and senior field engineer for nine western states for the National Electrical Manufacturers Association. He holds memberships in several professional organizations and has been an instructor at seminars on the National Electrical Code®.

Order form

Telephone orders: 715-247-5076 or 1-800-841-0383

Fax orders: 715-247-5062

Mail orders: Park Publishing, Inc., 350 Main Street
P.O. Box 387, Somerset, WI 54025-0387

E-mail: natlref@win.bright.net

Please send the following books:

_____ **Wiring Simplified** (ISBN 0-9603294-7-1)
$9.95 each + $3.00 shipping/handling = $12.95
2 copies, $7.96 each + $3.00 shipping/handling = $18.92

_____ **Practical Electrical Wiring** (ISBN 0-07-052395-9)
$39.95 each + $4.00 shipping/handling = $43.95

Sales tax: Minnesota residents add 6.5% sales tax

Prices subject to change without notice
Please inquire about quantity discounts

Name _____

Address _____

City _____ State _____ ZIP _____

Telephone (_____) _____
         area code

Payment must be submitted with order:

❑ Check or money order enclosed, payable to Park Publishing, Inc.

❑ VISA, MasterCard, or Discover credit card

Name on card _____

Card number _____ Exp. date ____ / ____

Cardholder signature _____